Lecture Notes in Mathematics

Edited by A. Dold and B. Eckmann

Subseries: Mathematisches Institut der Universität und
Max-Planck-Institut für Mathematik, Bonn

Adviser: F. Hirzebruch

1069

Matthias Kreck

Bordism of Diffeomorphisms and Related Topics

With an Appendix by Neal W. Stoltzfus

Springer-Verlag
Berlin Heidelberg New York Tokyo 1984

Author

Matthias Kreck
Fachbereich Mathematik der Universität Mainz
Saarstr. 21, 6500 Mainz, Federal Republic of Germany

AMS Subject Classification (1980): 57R50, 57R65, 57R90; 10C05, 57R15

ISBN 3-540-13362-3 Springer-Verlag Berlin Heidelberg New York Tokyo
ISBN 0-387-13362-3 Springer-Verlag New York Heidelberg Berlin Tokyo

Printing and binding: Beltz Offsetdruck, Hemsbach/Bergstr.
2146/3140-543210

CONTENTS

Introduction

The main theme of this work is the computation of the bordism group of diffeomorphisms. The problem is the following: Given two diffeomorphisms $f_i: M_i \longrightarrow M_i$ on closed manifolds does there exist a diffeomorphism $F: W \longrightarrow W$ on a manifold W with boundary $M_o + (-M_1)$ such that $F|_{\partial W} = f_o + f_1$. If we collect all diffeomorphism on m-dimensional manifolds and introduce the bordism relation as above we obtain the bordism group of diffeomorphisms. One can consider this group for manifolds with various structures but we concentrate for the moment on oriented manifolds. The bordism group of orientation preserving diffeomorphisms is denoted by Δ_m.

To describe the computation of the bordism group we introduce the following invariants. The most obvious invariants are the bordism class of the underlying manifolds and the bordism class of the mapping torus $M_f = \mathbb{R} \times_{\mathbb{Z}} M$ where the $\mathbb{Z}$-action on M is given by the diffeomorphism f.

For diffeomorphisms on even-dimensional manifolds there is a very interesting invariant called the isometric structure. It resides in the Witt group $W_\epsilon(\mathbb{Z};\mathbb{Z})$ of isometries of ϵ-symmetric unimodular bilinear forms on free finitely generated $\mathbb{Z}$-modules. The isometric structure of a diffeomorphism $f: M^{2n} \longrightarrow M^{2n}$ is defined by the triple $[H_n(M)/_{Tor}, o, f_*] \in W_{(-1)^n}(\mathbb{Z};\mathbb{Z})$ where o is the intersection form on M. We denote the isometric structure by I(M,f). Its equivalence class in the Witt group is a bordism invariant.

These invariants have been previously investigated by several people. For instance, it was known that the image of the mapping torus invariant $\Delta_m \longrightarrow \Omega_{m+1}$ is the kernel of the signature which we denote by $\hat{\Omega}_{m+1}$ ([34] , [51]). Furthermore, the isometric structure $I:\Delta_{2n} \longrightarrow W_{(-1)^n}(\mathbb{Z};\mathbb{Z})$ is surjective ([35] , Theorem A 5, p. 76).

Finnaly the Witt group $W_\epsilon(\mathbb{Z};\mathbb{Z})$ was investigated in ([35] , [42]). For instance one knows that $W_\epsilon(\mathbb{Z};\mathbb{Z})$ is isomorphic to $\mathbb{Z}^\infty \times \mathbb{Z}_2^\infty \times \mathbb{Z}_4^\infty$, the torsion free part is detected by equivariant signatures and the torsion is closely related to number theoretic invariants ([35] , [42]). The Witt group is related to a Witt group used in the computation of bordism of knots ([28] , [22]).

The computation of $W_\epsilon(\mathbb{Z};\mathbb{Z})$ has various consequences for Δ_{2n}. For instance it implies that Δ_{2n} is not finitely generated, a fact that first was obtained in ([51] , [30]), and moreover that $\Delta_{2n} \otimes \mathbb{Q}$ is not finite dimensional ([35]).

The main problem I started to investigate in 1975 was the computation of the kernel of the map given by the three invariants and to determine the relations between the invariants. The first result is that the kernel is trivial, yielding the amazing fact that a bordism class of a diffeomorphism is determined by those three invariants. The determination of the relations between the invariants was rather difficult. Besides the obvious fact that the signature of the underlying manifold is the signature of the form of the isometric structure there is a second relation possible between the isometric structure and the de Rham invariant of the mapping torus which is given by the Stiefel Whitney number $w_2 \cdot w_n(M_f)$ where $n = 3 \bmod 4$. Again the invariant can be expressed in terms of the isometric structure (Lemma 4.4). Together this gives a complete description of Δ_m for $m \geq 4$ (Theorem 5.7).

One can extend my proof to the case m = 3, too. This special case was first solved by P. Melvin [31]. For m = 2 my methods don't work as one expects if surgery methods are involved. In fact it was shown by A. Casson in 1979 (unpublished) that $I: \Delta_2 \longrightarrow W_-(\mathbb{Z};\mathbb{Z})$ is not injective. Soon after this Scharlemann determined the subgroup of Δ_2 generated by diffeomorphisms of the torus [40] and then Bonahon showed that $\Delta_2 \cong \mathbb{Z}^\infty \times \mathbb{Z}_2^\infty$ [6]. So, in contrast to the higher dimension it contains no 4-torsion. The same result was proved a little bit later in [17].

As mentioned above it is natural to consider bordism of diffeomorphisms on manifolds with various additional structures: for instance Spin structures, stably almost complex structures or framings. These structures are all special cases of so-called (B,f)-structures which were introduced in ([27], see also [43]). In this case we require that the diffeomorphism should preserve the (B,f)-structure on M and, to have a bordism relation for those diffeomorphisms, we must fix a homotopy between the original (B,f)-structure and the one induced by the diffeomorphisms. We only consider (B,f)-structures with B 1 - connected and denote the corresponding bordism group of diffeomorphisms by $\Delta_m^{(B,f)}$. It turns aut that the methods used for the computations of Δ_m extend to $\Delta_m^{(B,f)}$ and the result which includes the original result for Δ_m as a special case is the following.

<u>Theorem 9.9:</u> Let $m \geq 3$ and B be 1 - connected.

For m odd there is an isomorphism

$$\Delta_m^{(B,f)} \longrightarrow \Omega_m^{(B,f)} \oplus \hat{\Omega}_{m+1}^{(B,f)}, \quad \hat{\Omega}_{m+1}^{(B,f)} \text{ the kernel of the signature.}$$

For m = 2(4) there is a surjective map

$\Delta_m^{(B,f)} \longrightarrow \Omega_m^{(B,f)} \oplus \Omega_{m+1}^{(B,f)} \oplus W_-(\mathbb{Z};\mathbb{Z})$ with kernel a subgroup of $\mathbb{Z}/\tau(B,m+2)\mathbb{Z}$ where $\tau(B,m)$ is the smallest positive signature of a m-dimensional B-manifol

For m = 0(4) there is an exact sequence

$$0 \longrightarrow \Delta_m^{(B,f)} \longrightarrow W_+(\mathbb{Z};\mathbb{Z}) \oplus \Omega_m^{(B,f)} \oplus \Omega_{m+1}^{(B,f)} \to \mathbb{Z} \oplus \mathbb{Z}_2 \longrightarrow 0 \quad .$$

There are some other types of bordism groups of diffeomorphisms one could consider. For example orientation reversing diffeomorphisms on oriented manifolds, which we denote by Δ_m^-. In this case the bordism class of a diffeomorphism is determined by two invariants. The first lies in $\Omega_{n+1}(pt;\mathbb{Z}_2)$ the oriented bordism with $\mathbb{Z}_2$-coefficients which geometrically can be described as the bordism group of oriented manifolds whose boundary consists of two parts and an orientation reversing diffeomorphism from one part to the other. The map $\Delta_m^- \longrightarrow \Omega_{m+1}(pt;\mathbb{Z}_2)$ is given by $(M,f) \longmapsto M \times I$ with f as the orientation reversing diffeomorphism of the two parts of the boundary. The second invariant is a Witt group invariant sitting in the Witt group $W_\epsilon^-(\mathbb{Z};\mathbb{Z})$ consisting of ϵ-symmetric unimodular bilinear forms and an anti-isometry h, that is, $\langle v,w\rangle = -\langle hv,hw\rangle$. This Witt group is isomorphic to $\mathbb{Z}_2^\infty$ and it is analysed in ([50]). The computation of Δ_m^- can be expressed as follows:

<u>Theorem 9.17:</u> For m > 1 there are isomorphisms

$$\Delta_{4m-1}^- \longrightarrow \hat{\Omega}_{4m}(pt;\mathbb{Z}_2)$$

$$\Delta_{4m+1}^- \longrightarrow \Omega_{4m+2}(pt;\mathbb{Z}_2)$$

$$\Delta_{4m+2}^- \longrightarrow W_-^-(\mathbb{Z};\mathbb{Z}) \oplus \Omega_{4m+3}(pt;\mathbb{Z}_2)$$

and we have an injection

$$\Delta^-_{4m} \hookrightarrow W^-_+(\mathbb{Z};\mathbb{Z}) \oplus \hat{\Omega}_{4m+1}(pt;\mathbb{Z}_2) \text{ with cokernel } \mathbb{Z}_2 \text{ or } \{0\}.$$

Here $\hat{\Omega}_{4m}(pt;\mathbb{Z}_2)$ is the kernel of the signature mod 2.

Concerning the last map in Theorem 9.17 we know that $\Delta^-_{4m} \longrightarrow \Omega_{4m+1}(pt;\mathbb{Z}_2)$ is surjective. This fact, together with the surjectivity of the other homomorphisms in Theorem 9.17, has the following consequence. An oriented manifold is called reversible if it admits an orientation reversing diffeomorphism.

Corollary: A closed oriented manifold is bordant to a reversible manifold if and only if it has order 2 in Ω_*.

There is one other case of interest, diffeomorphisms of unoriented manifolds. It seems likely that one can do this case with the same methods, but the result is already contained in a paper of Quinn [38]. For completeness we state the result at the end of § 9.

The bordism groups of diffeomorphisms also have a graded ring structure induced by Cartesian product. An explicit set of generators is constructed in § 11.

There are two obvious other interpretations of bordism groups of diffeomorphisms. The first is bordism of $\mathbb{Z}$-actions on closed manifolds. This is related to bordism of diffeomorphism by assigning to a $\mathbb{Z}$-action on M the diffeomorphism corresponding to $1 \in \mathbb{Z}$. The other interpretation is the bordism group of differentiable fibre bundles over S^1 with m+1-dimensional total space. The correspondence between Δ_m and this group is given the mapping torus.

This last interpretation of Δ_m shows that the computation of Δ_m is closely related to the solution of the following (somewhat stronger) problem:

Let N be a m+2-dimensional manifold such that ∂N is the total space of a differentiable fibre bundle over S^1 and the projection $\partial N \longrightarrow S^1$ extends to a contin ous map p: $N \longrightarrow S^1$. Under what conditions is N bordant modulo boundary to a differentiable fibre bundle over S^1 extending the fibre bundle structure on the boundary?

We give a complete answer to this problem for (B,f)-manifolds (B as in 9.9), $m \geq 3$. To formulate the answer we need the following invariant of the boundary of N. Let ∂N be the mapping tours M_f and dim M = 4k+2. Then we denote the signature of the symmetric bilinear form on $H_{2k+1}(M;\mathbb{Q})$ given by $(x,y) \longmapsto (f_* - f_*^{-1})(x) \circ y$ by $\varphi(M,f)$. This invariant was introduced by W. Neumann and he showed that the signature of N = $\varphi(M,f)$ if the fibre bundle structure on ∂N extends to N [36].

This invariant and the isometric structure I(M,f) are the only obstruction to replacing N within its bordism class modulo boundary by a differentiable fibre bundle if B is 1 - connected and dim N > 4 (Theorem 9.1).

If $\partial N = \emptyset$ Theorem 9.1 implies that a closed (B,f)-manifold (B 1-connected) i bordant to a differentiable fibre bundle if and only if its signature vanishes. Thus we get a new proof of the corresponding result for oriented manifolds mentioned already before. This result was also known for unitary manifolds ([4]) but in general it seems to be new.

The problem of which closed manifolds are bordant to a differentiable fibre bundle is closely related to the computation of SK-groups: the Grothendieck group of all closed manifolds modulo the cutting and posting relations ([21] , [4]). For multiplicative (B,f)-structures we can completely identify the relation (Lemma 10.1) and, as a consequence of Theorem 9.1, we obtain the following result.

Theorem 10.2: Let (B,f) be multiplicative, B 1-connected, all dimensions >4.

Then: $SK_{2n+1}^{(B,f)} = \{0\}$, $SK_{4n+2}^{(B,f)} \cong \mathbb{Z}$ and

$$SK_{4n}^{(B,f)} \cong \begin{cases} \mathbb{Z} \oplus \mathbb{Z} & \text{if there exists a (B,f)-manifold with non-trivial signature} \\ \mathbb{Z} & \text{otherwise.} \end{cases}$$

Another type of SK-groups are those for manifolds with group actions. The case of compact groups actions is studied in [21] . For $\mathbb{Z}$-actions we can compute the group $\overline{SK}_{(\mathbb{Z},n)}$ which is obtained fron $SK_{(\mathbb{Z},n)}$, the SK-group of oriented $\mathbb{Z}$-actions, by dividing out zero bordant actions. This also works for unitary manifolds and the result is the following:

Theorem 10.4: For oriented or unitary manifolds with $\mathbb{Z}$-action of dim ≥ 3 we have:

$$\overline{SK}_{(\mathbb{Z},2n+1)} = 0 \quad \text{and} \quad \overline{SK}_{(\mathbb{Z},2n)} = W_{(-1)^n}(\mathbb{Z};\mathbb{Z}).$$

Concerning the proofs, the main step is the proof of Theorem 9.1. The idea is the following. Let $x \in S^1$ be a regular point of $p: N \longrightarrow S^1$. Then $F = p^{-1}(x)$ is a 2-sided codimension 1 manifold along which we can cut N up to obtain a manifold denoted by N_F. Then ∂N_F is a twisted double of F, twisted by f where f is the classifying diffeomorphism of the differentiable fibre bundle $\partial N \longrightarrow S^1$. The basic criterion we use is the following (which comes from the relative h-cobordism Theorem ([41])):

If F is 1-connected and N_F is a relative h-cobordism then the fibration on ∂N can be extended to N (Proposition 5.12).

To apply this criterion we modify an arbitrary N_F by certain surgical modifications, the addition and subtraction of handles, to obtain an h-cobordism. With the correct sequence of modifications it is not difficult to kill the homotopy groups of (N_F, F_i) below the middle dimension and, for odd-dimensional N_F, we can kill the middle dimensional homotopy groups using arguments modelled on those in Kervaire and Milnor's paper on homotopy spheres [23] .

The even dimensional case is much more difficult. Originally I considered the set of bordism classes of all such N_F and tried to find obstructions to replace N_F by the modifications above by an h-cobordism. When I found such obstructions I tried to find out the relation between them and the isometric structure. It turned out that under certain conditions on N_F the vanishing of the isometric structure implied the vanishing of these obstructions.

The additional conditions on N_F are too long to be stated here but the nice thing with them is that they can be achieved by an appropriate sequence of addition of handles (Proposition 8.3). Thus it was not necessary to introduce theses obstructions and I could directly prove that if the isometric structure vanishes and the other conditions of Theorem 9.1 are fulfilled then N is bordant modulo boundary to a differentiable fibre bundle.

I obtained this result in the oriented case in autumn 1975 and it was the main part of my Habilitationsschrift at the University of Bonn in 1976 [26]. An announcement of the computation of Δ_m appeared in the Bull.A.M.S. in 1976 [25] . I noticed rather soon after that that one can extend the proof to manifolds with additional structure. But I checked the details of this only in 1980. Another project I worked on was the extension of my results to diffeomorphism of singular manifolds in a topological space X. Before I finished that I heard of a paper of F. Quinn which contains these results [38] . Especially this paper contains another computation of Δ_m in terms of an exact sequence.

All terms of this exact sequence are obviously related to the above invariants except that instead of the Witt group of isometric structures a Witt group of bilinear forms (without any symmetry) $W^S(\mathbb{Z})$ ($A(\mathbb{Z})$ in our notation) occured. The relation between the two Witt groups is not discussed in Quinn's paper.

I discussed this problem with Neal Stolzfus in 1979/80 and we thought that its solution and some other algebraic results should be contained in these lecture notes. This appendix contains the explicit algebraic and geometric relationship between the two Witt groups which has the following nice consequences. First, the geometric de Rham invariant of a mapping torus is determined by the isometric structure (Lemma 4.4). The original proof of Lemma 4.4 in my Habilitationsschrift is not correct as was pointed out to me by W. Neumann. I then found a correct but lengthy proof which I will not give and will instead refer to the appendix.

Another corollary of the exact sequence relating the two Witt groups is a proof of Proposition 9.8 which determins the quotient of $W_+(\mathbb{Z};\mathbb{Z})$ by the subgroup of isometric structures of even forms. This result is necessary for the computation of bordism of diffeomorphism of arbitrary (B,f)-manifolds.

The results in the appendix take the following form: First, it is necessary to find a natural home for Quinn's invariant which measures the obstruction to extending an open book decomposition on the boundary of a manifold to the interior and to generalize Quinn's geometric exact sequences to inculde this group.

We then use Quinn's invariant and the isometric structure to give a natural transformation to the following exact sequence:

Theorem 1 (Appendix) $0 \longrightarrow W_{-\epsilon}(\mathbb{Z}) \rightarrow A(\mathbb{Z}) \longrightarrow W_{\epsilon}(\mathbb{Z};\mathbb{Z}) \longrightarrow W_{\epsilon}(\mathbb{Z}) \oplus$ Coker $\partial_{-\epsilon} \longrightarrow 0$ is exact.

Here $A(\mathbb{Z})$ (denoted $W^S(\mathbb{Z})$ by Quinn) is the Witt group of non-singular bilinear (possibly without symmetry, that is, asymmetric and hence the notation) on f. g. $\mathbb{Z}$-modules. $W_{\epsilon}(\mathbb{Z})$ is the Witt group of ϵ-symmetric forms over $\mathbb{Z}$ which are well known to be: $W_{+1}(\mathbb{Z}) = \mathbb{Z}$, $W_{-1}(\mathbb{Z}) = 0$ and $W_{\epsilon}(\mathbb{Z}, \mathbb{Z})$ is the Witt group of ϵ-isometric structures. The mapping connecting the two is a "bilinearization map" which generalizes the well known connection between the Seifert matrix of a fibred knot and its monodromy. The Coker $\partial_{-\epsilon}$ is the cokernel of the boundary in the localization sequence

$$W_{\epsilon}(\mathbb{Z}) \longrightarrow W_{\epsilon}(\mathbb{Q}) \xrightarrow{\partial_{\epsilon}} W_{\epsilon}(\mathbb{Q}/\mathbb{Z})$$

$$\text{Coker } \partial_{\epsilon} = \begin{cases} 0 & \epsilon = 1 \\ \mathbb{Z}_2 & \epsilon = -1 \end{cases}$$

with the non-trivial element related to the de Rham invariant. From the computations of the other terms, we see that the infinitely generated groups of Witt classes of asymmetric and isometric structures are almost the same.

I would like to make the following acknowledgments. When I wrote my Habilitationsschrift in Bonn I had numerous and helpful discussions with Karl-Heinz Knapp, Walter D. Neumann and Erich Ossa. Parts of the manuscript were written up when I visited the Matematisk Institut in Arhus in 1978 and the IHES in Bures in 1979. My special thank goes to Neal Stoltzfus for several fruitful discussions, for writing the appendix and for helping me with the translation. Neal Stoltzfus has been partially supported by the National Science Foundation and the Institute for Advanced Study during the writing of the appendix.

Finally I would like to thank Frau Pahnke and Frau Schack for a careful typing of the manuscript.

§ 1 Bordism groups of orientation preserving diffeomorphisms.

We consider diffeomorphisms of C^∞-manifolds. In the following we assume that all manifolds are compact and oriented and all diffeomorphisms are C^∞-differentiable and orientation preserving. A pair (M,f) consisting of a n-dimensional manifold M and a diffeomorphism $f : M \longrightarrow M$ is denoted as a n-dimensional diffeomorphism. Two diffeomorphisms (M_1,f_1) and (M_2,f_2) are called equivalent if there exists a diffeomorphism $g : M_1 \longrightarrow M_2$ such that the diagram

$$\begin{array}{ccc} M_1 & \xrightarrow{f_1} & M_1 \\ \Big\downarrow g & & \Big\downarrow g \\ M_2 & \xrightarrow[f_2]{} & M_2 \end{array}$$

commutes. Especially two diffeomorphisms of the same manifold are equivalent if and only if they are conjugate.

If M has a boundary we denote $(\partial M, f|_{\partial M})$ by $\partial(M,f)$. The sum of two diffeomorphisms (M_1,f_1) and (M_2,f_2) is given by the disjoint union M_1+M_2 and the diffeomorphism on M_1+M_2 induced by f_1 and f_2. We denote this sum by $(M_1,f_1) + (M_2,f_2)$. If we change the orientation on M we denote the corresponding diffeomorphism by $-(M,f)$. For $(M_1,f_2) + (-(M_2,f_2))$ we write $(M_1,f_1) - (M_2,f_2)$.

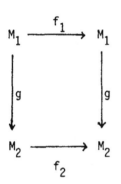

Definition 1.1: Two n-dimensional diffeomorphisms (M_1,f_1) and (M_2,f_2) of closed manifolds are called bordant (notation: $M_1,f_1) \sim (M_2,f_2)$ if there exists a diffeomorphism (N,g) of a n+1-dimensional manifold with boundary

such that $\partial(N,g) = (M_1,f_1) - (M_2,f_2)$.

One obtains very special cases of bordant diffeomorphisms if they are (pseudo-) isotopic, where $f_1,f_2 : M \longrightarrow M$ are called (pseudo-) isotopic if there exists a diffeomorphism $F : M \times I \longrightarrow M \times I$ extending f_1 and f_2 and which for an isotopy has to commute with the projection onto I.

Bordism of diffeomorphisms is an equivalence relation. Symmetry and reflexivity are obvious, the transitivity is a consequence of the following Lemma.

Lemma 1.2: Let (M,f) be a diffeomorphism and V a union of boundary components of M which is mapped to itself by f. Then f is isotopic to a diffeomorphism g which is equal to f on V and outside a small neighbourhood of V and which is a product with Id near V, i.e. on a collar of V in M of the form $f|_V \times \mathrm{Id}$.

The proof of this Lemma is a consequence of the uniqueness theorem for a collar ([7] , p.141).

Corollary 1.3: Let (M_1,f_1) and (M_2,f_2) be diffeomorphisms which are equal on a common union of boundary components V. Then there exists a diffeomorphism on $M_1 \cup_V -M_2$ which is equal to $f_1 \cup f_2$ outside a small neighbourhood of V in $M_1 \cup_V -M_2$.

This corollary implies the transitivity of the bordism relation.

Definition 1.4: We denote the bordism class of a diffeomorphism (M,f) by $[M,f]$. The set of n-dimensional bordism classes of diffeomorphisms is an abelian group under disjoint union and we denote it by Δ_n. There is a graded ring structure on $\Delta_* := \bigoplus_n \Delta_n$ given by the cartesian product of diffeomorphisms.

Remark 1.5: There are two other interpretations of these bordism groups.
1.) For a compact Lie group G there is defined the bordism group $O_*(G)$ of G-operations on manifolds (compare [1], p.57, the notation there is $\Omega_*(G,G)$). It consists of bordism classes of orientation preserving differentiable G-actions on closed oriented manifolds. This definition can be extended to $G = \mathbb{Z}$. The only difficult point to define $O_*(G)$ for non-compact groups too is the equivariant collaring theorem. But for $G = \mathbb{Z}$ this follows from Lemma 1.2. It is obvious that $O_*(\mathbb{Z}) \cong \Delta_*$ the isomorphism maps a $\mathbb{Z}$-operation to the diffeomorphism given by $1 \in \mathbb{Z}$. There is known something about $O_*(G)$ for a compact Lie group. Our computation of Δ_* is the first computation of bordism groups of G-operations for non-compact G.

2.) An other interpretation of Δ_* is given by bordism of differentiable fibre bundles over S^1. It consists of m-dimensional differentiable oriented fibre bundles with closed fibre M over S^1 modulo those bounding a m+1-dimensional oriented differentiable fibre bundle over S^1 with compact fibre F such that $\partial F = M$. We denote the corresponding bordism group by $\mathfrak{F}_m$ where m is the dimension of the total space. The mapping torus M_f of a diffeomorphism (M,f), is defined as $I \times M/(o,x) \sim (1,f(x)) = \mathbb{R} \times_{\mathbb{Z}} M$ where $\mathbb{Z}$ operates on $\mathbb{R}$ by translation and on M by f^n. The assignment $[M,f] \longmapsto [p:M_f \longrightarrow S^1]$ induces an isomorphism $\Delta_m \longrightarrow \mathfrak{F}_{m+1}$.

Now, we want to show that every diffeomorphism is bordant to a diffeomorphism on a connected manifold. This is based on the fact that every diffeomorphism of a connected manifold is isotopic to a diffeomorphism leaving an embedded disk fixed.

Lemma 1.6: Let (M^n,f) be a diffeomorphism of a connected manifold. Then, for every embedding $i : \mathbb{R}^n \longrightarrow M$ there exists a diffeomorphism g isotipic to f such that $g|_{i(D^n)}$ is the identity where $D^n = \{x \in \mathbb{R}^n \mid \|x\| \leq 1\}$.

The proof is a consequence of the uniqueness theorem for embeddings $\mathbb{R}^n \longrightarrow M$ ([7] , p.101).

As a consequence, for diffeomorphisms $(M_1{}^n,f_1)$ and $(M_2{}^n,f_2)$ on connected manifolds we can define a connected sum $(M_1 \# M_2, f_1 \# f_2)$ which is well defined up to isotopy of diffeomorphisms. It is obvious, that $(M_1 \# M_2, f_1 \# f_2)$ is bordant to $(M_1,f_1) + (M_2,f_2)$, a bordism is given by $M_1 \times I \natural M_2 \times I$ where the boundary connected sum is taken within $M_1 \times \{1\}$ and $M_2 \times \{1\}$ and the diffeomorphism on it is $f_1 \natural f_2$.

Lemma 1.7: Every diffeomorphism of a positive-dimensional manifold is bordant to a diffeomorphism of a connected manifold.

Proof: Let (M^n,f) be a diffeomorphism. If there are two components of M which are mapped to themselves by f we can form the connected sum of these components to get a diffeomorphism on a manifold with fewer components than M.

Now, let N be a component of M with $f(N) \neq N$. As M is compact it has a finite number of components and there is a smallest positive number r with $f^r(N) = N$. We consider the unit ball D^{n+1} as a subspace of $\mathbb{C} \times \mathbb{R}^{n-1}$. Let λ be a primitive r-th root of unity and h the diffeomorphism on D^{n+1} mapping $(z,x_1,\ldots,x_{n-1})$ to $(\lambda \cdot z,x_1,\ldots,x_{n-1})$. Now, we choose an embedding $i: D^n \longrightarrow S^n \subset \mathbb{C} \times \mathbb{R}^{n-1}$ such that $i(D^n)$, $h \circ i(D^n),\ldots,$ $h^{r-1} \circ i(D^n)$ are disjoint and an embedding j of D^n into V. Then we form a manifold W out of M x I and D^{n+1} identifying $f^k \circ j(D^n)$ in $M \times \{1\}$, with $h^k \circ i(D^n)$ in S^n and we remove the resulting corners (compare [13], I.3). f x Id and h induce a diffeomorphism on W. The boundary of W consists of the complement of $\bigcup_{i=1}^{r} f^i(N)$ and of $\#_r N$. Thus we get a bordism between (M,f) and a diffeomorphism on a manifold with fewer components than M. By induction we can make (M,f) bordant to a diffeomorphism on a connected manifold.

q.e.d.

As a consequence we see that every diffeomorphism on a 1-dimensional manifold is bordant to a diffeomorphism on S^1. As every diffeomorphism on S^1 can be extended to D^2 this implies:

<u>Corollary 1.8:</u> Δ_1 is the trivial group.

§ 2 Report about equivariant Witt groups

In this chapter we want to define Witt groups of isometric structures and give a report about the results which are important in our context. These groups play an important role for the computation of Δ_{2m}. They also occur at the computation of bordism groups of knots. The main part of the computation of the Witt groups can be found in Kervaire's paper [22]. These computations were extended by W.D. Neumann [35]. This paper is the basis for our report. Similar computations can be found in [42].

Definition 2.1: For $R = \mathbb{Z}$ or a field an ϵ-isometric structure over R is a triple (V,s,f), where V is a free finite-dimensional R-module, $s : V \times V \longrightarrow R$ a symmetric ($\epsilon = +1$) (antisymmetric ($\epsilon = -1$)) unimodular bilinear form and $f : V \longrightarrow V$ an isometry of (V,s) into itself, i.e. for all $x,y \in V$ is $s(x,y) = s(f(x), f(y))$.

The sum of two ϵ-isometric structures is defined by the orthogonal direct sum: $(V_1,s_1,f_1) + (V_2,s_2,f_2) = (V_1 \oplus V_2, s_1 \oplus s_2, f_1 \oplus f_2)$.

An ϵ-isometric structure is called metabolic or null-bordant if there exists an invariant subkernel, i.e. a submodule $W \subset V$ with the properties $f(W) = W$ and $W = W^{\perp}$.

Two ϵ-isometric structures (V_1,s_1,f_1) and (V_2,s_2,f_2) are called bordant if $(V_1,s_1,f_1) + (V_2,-s_2,f_2)$ is null-bordant. This is an equivalence relation.

The equivalence classes $[V,s,f]$ of ϵ-isometric structures over R form an abelian group. The inverse element of $[V,s,f]$ is $[V,-s,f]$. This group is denoted by $W_\epsilon(\mathbb{Z};R)$, the Witt group of ϵ-isometric structures over R.

The tensor product defines a product $W_{\epsilon_1}(\mathbb{Z};R) \otimes W_{\epsilon_2}(\mathbb{Z};R) \longrightarrow W_{\epsilon_1 \cdot \epsilon_2}(\mathbb{Z};R)$. For special reasons we replace this map by its negative for $\epsilon_1 = \epsilon_2 = -1$. $W_*(\mathbb{Z};R) := W_+(\mathbb{Z};R) \oplus W_-(\mathbb{Z};R)$ is made into a $\mathbb{Z}_2$-graded ring by this product.

Analogously we define the Witt group of isometries of hermitian vector spaces over $\mathbb{C}$, denoted by $W\,U(\mathbb{Z};\mathbb{C})$. $W\,U(\mathbb{Z};\mathbb{C})$ is a ring under tensor product.

Remarks 2.2: 1.) Instead of the property $f(W) = W$ and $W = W^\perp$ it is sufficient to require $f(W) = W$, $W \subset W^\perp$ and $2 \cdot \dim W = \dim V$.

2.) The transitivity of the bordism relations is not completely obvious (compare [35], Lemma 2.1).

3.) An ϵ-isometric structure is an isometry of (V,s). This isometry corresponds to a representation of $\mathbb{Z}$ on (V,s). If we consider more generally representations of a group G on (V,s) and introduce the corresponding Witt group, we obtain the group $W_\epsilon(G;R)$ in Neumann's notation. This clarifies the notation of $W_\epsilon(\mathbb{Z};R)$.

For the computation of Δ_{2m} we need the Witt group $W_\epsilon(\mathbb{Z};\mathbb{Z})$. This is closely related to the Witt group $W_\epsilon(\mathbb{Z};\mathbb{Q})$. For tensoring the underlying module with $\mathbb{Q}$ gives an embedding of $W_\epsilon(\mathbb{Z};\mathbb{Z})$ into $W_\epsilon(\mathbb{Z};\mathbb{Q})$. As mentioned before, similar groups occur at the computation of concordance of knots and were computed in this context.

Theorem 2.3 ([28] , [22]): $W_\epsilon(\mathbb{Z};\mathbb{Q}) \cong \mathbb{Z}^\infty \oplus \mathbb{Z}_2^\infty \oplus \mathbb{Z}_4^\infty$

To understand the group $W_\epsilon(\mathbb{Z};\mathbb{Q})$ and also $W_\epsilon(\mathbb{Z};\mathbb{Z})$ it is useful to have a system of numerical invariants which classify the elements in $W_\epsilon(\mathbb{Z};\mathbb{Q})$. Neumann has found such a system for $W_\epsilon(\mathbb{Z};\mathbb{Q}) \otimes \mathbb{Q}$ and we want to describe it here.

For this we first describe the structure of $W\,U(\mathbb{Z})$.

Theorem 2.4 ([35] , § 9): The map assigning to $\sum_i n_i x_i \in \mathbb{Z}[S^1]$ the class $\sum [\mathbb{C}^{|n_i|}, \operatorname{sign}(n_i)\cdot t, x_i \cdot \mathrm{Id}]$ in $W\,U(\mathbb{Z})$ is a ring isomorphism $\mathbb{Z}[S^1] \longrightarrow W\,U(\mathbb{Z})$.
Here t is the standard hermitian form on $\mathbb{C}^n$.

Now, we define the equivariant signature of elements in $W\,U(\mathbb{Z})$.

Definition 2.5: Write $[V,s,f] \in W\,U(\mathbb{Z})$ as $\sum_i [\mathbb{C}^{|n_i|}, \operatorname{sign}(n_i)\cdot t, x_i \cdot \mathrm{Id}]$. Then the equivariant signature is defined as

$$\operatorname{sign}[V,s,f] := \sum_{\substack{i \\ \operatorname{sign} n_i > 0}} |n_i| \cdot x_i - \sum_{\substack{i \\ \operatorname{sign} n_i < 0}} |n_i| \cdot x_i$$

This equivariant signature can also be described in the following way: In $W\,U(\mathbb{Z})$ we can split $[V,s,f]$ as $[V_1,s_1,f_1] + [V_2,s_2,f_2]$ with s_1 positive definite and s_2 negative definite. Then

$$\operatorname{sign}[V,s,f] = \operatorname{trace} f_1 - \operatorname{trace} f_2$$

<u>Proposition 2.6:</u> The equivariant signature has the following properties:

1.) sign $([V_1,s_1,f_1]+[V_2,s_2,f_2]) = \text{sign}\,[V_1,s_1,f_1]+\text{sign}\,[V_2,s_2,f_2]$

2.) sign $([V_1,s_1,f_1]\otimes[V_2,s_2,f_2]) = \text{sign}\,[V_1,s_1,f_1]\cdot\text{sign}\,[V_2,s_2,f_2]$
Thus, the equivariant signature is a ring homomorphism $W\,U\,(\mathbb{Z}) \longrightarrow \mathbb{C}$.

3.) ([35], Theorem 9.3) If sign $[V,s,f^n]$ vanishes for all $n \in \mathbb{Z}$ then $[V,s,f] = 0$.
Thus sign $[V,s,f^n]$ $(n \in \mathbb{Z})$ classifies elements in $W\,U\,(\mathbb{Z})$.

Next we want to introduce the equivariant signature of an ϵ-isometric structure over $R = \mathbb{Z}$ or $\mathbb{Q}$. For this we consider the following map $\emptyset : W_\epsilon(\mathbb{Z};R) \longrightarrow W\,U\,(\mathbb{Z})$, defined as $\emptyset\,[V,s,f] = [V\otimes\mathbb{C},s,f]$ for $\epsilon=+1$ and as $\emptyset\,[V,s,f] = [V\otimes\mathbb{C},i\cdot s,f]$ for $\epsilon=-1$. Here s resp. $i\cdot s$ and f stand for the unique extension of the real resp. imaginar form and linear map from V to $V\otimes\mathbb{C}$. The ϵ-symmetry of the original form implies that the form on $V\otimes\mathbb{C}$ is hermitian. With the sign convention for the product in $W_*(\mathbb{Z};R)$ defined above this map is a ring homomorphism. Neumann shows that the kernel of $\emptyset$ consists of torsion only.

<u>Definition 2.7:</u> The equivariant signature of $[V,s,f] \in W_\epsilon(\mathbb{Z};R)$ is defined as sign $(\emptyset\,[V,s,f])$, again denoted as sign $[V,s,f]$. If $\epsilon = +1$ this equivariant signature is real, if $\epsilon = -1$ purely imaginar.

From this and the property 3 of Proposition 2.6 we can formulate the following result.

Theorem 2.8 [35] : For $R = \mathbb{Z}$ or $\mathbb{Q}$ elements $[V,s,f]$ in $W_\epsilon(\mathbb{Z};R) \otimes \mathbb{Q}$ are completely classified by sign $[V,s,f^n]$, $n \in \mathbb{Z}$.

As mentioned above there is no simple system of numerical invariants known which classifies the torsion of $W_\epsilon(\mathbb{Z};R)$. Medrano has introduced a torsion invariant which can be used to show the existence of infinitely many 2-torsion elements in $W_*(\mathbb{Z};\mathbb{Z})$ [30]. For completeness we repeat his definition here and refer to [30] for the details.

We consider polynomials $F(t)$ over $\mathbb{Z}$ with the properties $F(0) = \pm 1$ and $F(t) = \pm t^d F(t^{-1})$, where d is the degree of F. For these polynomials we introduce an equivalence relation as follows. $F_1(t)$ and $F_2(t)$ are equivalent if $F_1(t) \cdot F_2(t)$ can be written as $\pm t^k f(t) \cdot f(t^{-1})$ where $k = \text{grad } f$ and $2k = \text{grad } F_1 + \text{grad } F_2$. The set of equivalence classes forms a multiplicative group P. Every element of P has order 2.

For (V,s,f) an ϵ-isometric structure over $\mathbb{Z}$ or $\mathbb{Q}$ we denote the characteristic polynomial $\det(f - t\cdot \text{Id})$ by $p_f(t)$. As s is unimodular and f an isometry it follows that $p_f(t)$ fulfills the equation $p_f(t) = \pm t^d p_f(t^{-1})$. And $p_f(0) = \det f = \pm 1$.

It turns out that $[p_f] \in P$ vanishes for null-bordant isometric structures. Thus we get a homomorphism

$$p : W_\epsilon(\mathbb{Z};\mathbb{Z}) \longrightarrow P$$

$$[V,s,f] \longmapsto p_f \quad .$$

§ 3 The isometric structure of a diffeomorphism

In this chapter we want to continue our report and want to introduce invariants for bordism classes of diffeomorphisms. These invariants were first introduced by Medrano [30] and then investigated by Neumann ([35] , p.73 ff).

Let (M^{2n},f) be a diffeomorphism of a 2n-dimensional closed manifold. The intersection from s on $H_n\ (M;\mathbb{Z})/_{Tor}$ is $(-1)^n$-symmetric and unimodular by Poincaré duality. The diffeomorphism induces an isometry $f_* : H_n(M;\mathbb{Z})/_{Tor} \longrightarrow H_n(M,\mathbb{Z})/_{Tor}$.

Definition 3.1: The isometric structure I(M;f) of a diffeomorphism (M,f) is defined as $[H_n(M;\mathbb{Z})/_{Tor}, s, f_*] \in W_{(-1)^n}(\mathbb{Z};\mathbb{Z})$.
The equivariant signature of (M,f) is defined as sign I(M,f) and we write for this sign (M,f).
A third invariant is given by the composition of I and p as defined at the end of § 2. We denote this composition again by $p : \Delta_{2n} \longrightarrow P$.

Remark 3.2: If f is contained in a compact Lie group operating on M this definition of sign (M,f) coincides with the equivariant signature of Atiyah and Singer [4]. For $f_* = Id$ we get the Hirzebruch-signature of M.

The isometric structure has the following properties.

Lemma 3.3: a) If (M,f) is null-bordant then I(M,f) = 0.
b) $I((M_1,f_1) + (M_2,f_2)) = I(M_1,f_1) + I(M_2,f_2)$

$$I(M_1 \times M_2, f_1 \times f_2) = I(M_1,f_1) \cdot I(M_2,f_2).$$

c) If (M_1,f_1) and (M_2,f_2) are diffeomorphisms with boundary and $\varphi_1, \varphi_2 : \partial M_1 \longrightarrow \partial M_2$ are orientation reversing diffeomorphisms commuting with f_1 and f_2. Then

$$I(M_1 \underset{\varphi_1}{\cup} M_2, f_1 \cup f_2) = I(M_1 \underset{\varphi_2}{\cup} M_2, f_1 \cup f_2).$$

This is an analogous formula as the Novikov-additivity for the signature.

Lemma 3.3 implies that analogous statements hold for the equivariant signature and the map p. Thus our three invariants are homomorphisms.

The proof is nearly the same as in the case of diffeomorphisms of finite order. For instance if (M,f) bounds a diffeomorphism (N,g) then the kernel of $i_* : H_n(M;\mathbb{Z})/_{Tor} \longrightarrow H_n(N;\mathbb{Z})/_{Tor}$ is an invariant subkernel of I(M,f). Details can be found in ([35] , p.73 ff.).

Examples 3.4: The group of isotopy classes of orientation preserving diffeomorphisms on the torus $S^1 \times S^1$ is isomorphic to $SL(2;\mathbb{Z})$. The isometric structure of a diffeomorphism corresponding to $A \in SL(2;\mathbb{Z})$ is equal to $\left[\mathbb{Z}^2, \begin{pmatrix} 0 & 1 \\ -1 & 0 \end{pmatrix}, A \right]$. The equivariant signature of this diffeomorphism depends only on the trace of A and the values are as follows:

$$\text{trace } A = 0 \ : \ \text{sign } (S^1 \times S^1, A) = -2i$$

$$\text{trace } A = 1 \ : \ \text{sign } (S^1 \times S^1, A) = -\sqrt{3}\, i$$

$$\text{trace } A = -1 \ : \ \text{sign}\,(S^1 \times S^1, A) = -\sqrt{5}\,i$$

$$|\text{trace } A| \geq 2 : \ \text{sign}\,(S^1 \times S^1, A) = 0$$

This follows by a straightforward computation.

On the other hand it is easy to see that $I(S^1 \times S^1, A) = 0$ if and only if $|\text{trace } A| = 2$. In this case $(S^1 \times S^1, A)$ is null-bordant.

These computations and the fact that $W_-(\mathbb{Z};\mathbb{Z}) \otimes \mathbb{Q}$ is completely classified by equivariant signatures imply that $I(S^1 \times S^1, A)$ has infinite order for $|\text{trace } A| \leq 1$ and is a non-trivial torsion element for $[\text{trace } A] > 2$. But it is in general a difficult problem to determine the order of $[S^1 \times S^1, A] \in \Delta_2$ ([35] , Lemma 8.3).

For instance all diffeomorphisms corresponding to $A_r = \begin{pmatrix} r^2+1 & r \\ r & 1 \end{pmatrix}$, $r > 0$, have order 2 in Δ_2 as $\begin{pmatrix} r & 1 \\ 1 & 0 \end{pmatrix}$ gives an orientation reversing diffeomorphism commuting with A_r. In fact A_r is the square of this diffeomorphism. On the other hand it is easy to compute the invariant p for these diffeomorphisms and it turns out that it takes different values for all A_r. Thus, the A_r give us infinitely many diffeomorphisms or order 2 in Δ_2.

It is not so easy to construct diffeomorphisms on $S^1 \times S^1$ whose isometric structure has order 4 in $W_-(\mathbb{Z};\mathbb{Z})$. But Neumann has shown that there exist infinitely many diffeomorphisms of this type ([35] , p.59).

By products of these elements of order 2 and 4 in $W_-(\mathbb{Z};\mathbb{Z})$ we get infinitely many elements of order 2 and 4 in $W_+(\mathbb{Z};\mathbb{Z})$.

To show the existence of infinitely many linear independent diffeomorphisms in $\Delta_{2n} \otimes \mathbb{Q}$ it is enough to give infinitely many diffeomorphisms whose equivariant signatures are linear independent over $\mathbb{Z}$. But such a sequence can easily be found already for diffeomorphisms of finite order and one can use the Atiyah-Singer equivariant signature Theorem for the computation of these signatures. We leave this as an exercise to the reader.

As $W_\epsilon(\mathbb{Z};\mathbb{Z})$ can be embedded into $W_\epsilon(\mathbb{Z};\mathbb{Q}) \cong \mathbb{Z}^\infty \oplus \mathbb{Z}_2^\infty \oplus \mathbb{Z}_4^\infty$ these examples show the following result.

Theorem 3.5 ([35], p.58 and p.73 ff.): $W_\epsilon(\mathbb{Z};\mathbb{Z}) \cong \mathbb{Z}^\infty \oplus \mathbb{Z}_2^\infty \oplus \mathbb{Z}_4^\infty$. Δ_{2n} has infinitely many elements of order 2 and $\Delta_{2n} \otimes \mathbb{Q}$ is not finite-dimensional.

The fact that Δ_{2n} is not finitely generated was first shown by Winkelnkemper for n odd [51] and by Medrano for n even [30] using the characteristic polynomial of a diffeomorphism.

Now it is natural to ask whether the homomorphism $I: \Delta_{2n} \longrightarrow W_{(-1)^n}(\mathbb{Z};\mathbb{Z})$ is surjective. Neumann has shown that this is true for all $n > 0$. Because of the multiplicativity of the isometric structure it is enough to show this for n = 1 and 2. The general case follows by multiplication with $[P_{2K}\mathbb{C}, Id]$. The case n = 1 follows from theorems of Nielsen about diffeomorphisms on surfaces and the case n = 2 is based on Wall's result on diffeomorphisms on 1-connected 4-manifolds. Details can be found in ([35], p.76).

Theorem 3.6 [35] : The isometric structure gives a surjective homomorphism $I : \Delta_{2n} \longrightarrow W_{(-1)^n}(\mathbb{Z};\mathbb{Z})$ for all $n > 0$.

We end this chapter with the following remark concerning the equivariant signature.

Remark 3.7: The Atiyah-Singer equivariant signature Theorem expresses the equivariant signature of a diffeomorphism of finite order in terms of the normal budles of the fixed point manifolds [4]. Especially this implies that the equivariant signature vanishes if the operation is free. Such a result can't be true for diffeomorphisms of infinite order. For Neumann has given an example of a diffeomorphism (M,f) where f^n has no fixed point for all $n \neq 0$ but the equivariant signature is non-zero ([35] p.76).

§ 4 The mapping torus of a diffeomorphism

In the last chapter we have introduced invariants for diffeomorphisms of even-dimensional manifolds. In this chapter we study two other invariants, the mapping torus of a diffeomorphism and the bordism class of the underlying manifold.

The mapping torus M_f of a diffeomorphism (M,f) is defined as the identification space $I \times M/_{(0,x)\sim(1,f(x))} = \mathbb{R} \times_{\mathbb{Z}} M$. The orientations of M and $\mathbb{R}$ induce an orientation of M_f. If (N,g) is a bordism between (M,f) and (M',f') then N_g is bordism between M_f and $M'_{f'}$. By this we get a homomorphism

$$\Delta_m \longrightarrow \Omega_{m+1}, \quad [M,f] \longmapsto [M_f]$$

where Ω_{m+1} is the bordism group of closed oriented differentiable m+1-dimensional manifolds.

Remark 4.1: If f and g are diffeomorphisms of M then $[M_{f\cdot g}] = [M_f] + [M_g]$.

Several authors have investigated the question what the image of this map is. As the signature of the total space of a fibration over S^1 vanishes this image is contained in the kernel of $\tau : \Omega_{m+1} \longrightarrow \mathbb{Z}$ where τ is the Hirzebruch-signature. We denote the kernel of τ by $\hat{\Omega}_{m+1}$. Neumann [34] and Winkelnkemper [51] have independently shown that the image of our homomorphism is equal to this kernel.

Theorem 4.2 ([34], [51]): The image of

$$\Delta_m \longrightarrow \Omega_{m+1},\ [M,f] \longmapsto [M_f]$$

is equal to $\hat{\Omega}_{m+1}$, the kernel of the signature homomorphism.

Remark 4.3: We will give a new proof of this result later.

The second invariant is given by forgetting the diffeomorphism and considering only the bordism class of the underlying manifold. It is the homomorphism

$$\Delta_m \longrightarrow \Omega_m,\quad [M,f] \longmapsto [M]$$

In the following we want to study the relations between the invariants given by the isometric structure, the mapping torus and by the underlying bordism class.

For m odd we have only the last two invariants and it is obvious that the map

$$\theta_m: \Delta_m \longrightarrow \Omega_m \oplus \hat{\Omega}_{m+1},\ [M,f] \qquad ([M],[M_f])$$

is surjective.

For m even we have the three invariants. We first study the image of

$$\Delta_{2K} \longrightarrow W_{(-1)^k}(\mathbb{Z};\mathbb{Z}) \oplus \Omega_{2k},\ [M,f] \longmapsto (I(M,f),[M]).$$

For k odd this map is surjective. This follows for $k = 1$ from the surjectivity of I and the fact that $\Omega_2 = \{0\}$. For k odd >1 it follows from this using multiplication with $[P_{k-1}\mathbb{C}, \mathrm{Id}]$.

For k even the map is not surjective as the signature of M is equal to the signature of the bilinear form of the isometric structure I (M,f). But this is the only relation between I (M,f) and $[M]$. We formulate this as follows:

For k even the map $[M,f] \mapsto (I(M,f), [M])$ induces a surjective map $\Delta_{2k} \longrightarrow W_+(\mathbb{Z};\mathbb{Z})/\langle[\mathbb{Z},1,1]\rangle \oplus \Omega_{2k}$. As in the case of odd k it suffices to show surjectivity for k = 2.

To determine the image of the map

$\Delta_{2k} \longrightarrow W_{(-1)^k}(\mathbb{Z};\mathbb{Z}) \oplus \Omega_{2k} \oplus \Omega_{2k+1}$, $[M,f] \longmapsto (I(M,f), [M], [M_f])$

I first recall that Ω_{2k+1} consists only of elements of order 2 which are classified by their Stiefel-Whitney numbers. Thus one has to determine the relations between the Stiefel-Whitney numbers of the mapping torus and the isometric structure of a diffeomorphism. We will see that such a relation only exists for even k. Namely for even k we can determine the Stiefel-Whitney number $w_2(M_f) \cdot w_{2k-1}(M_f)$ from the isometric structure. By a Theorem of Lusztig, Milnor and Peterson this Stiefel-Whitney number is equal to the difference of semi characteristics with coefficients in $\mathbb{Q}$ and $\mathbb{Z}_2$ [29]:

$$\sum \text{rank } H_{2i}(M_f;\mathbb{Q}) - \sum \text{rank } H_{2i}(M_f;\mathbb{Z}_2),$$

which is also called the de Rham invariant of M_f (compare [3]).

<u>Lemma 4.4:</u> Let f be a diffeomorphism on a 4k - dim manifold M.

$$w_2(M_f) \cdot w_{4k-1}(M_f) = \dim \text{Ker}\,(1-H_{2k}(f,\mathbb{Q})) - \dim \text{Ker}\,(1-H_{2k}(f,\mathbb{Z}_2)) \bmod 2 =$$

$$= \dim \text{Ker}\,(1-H_{2k}(f;\mathbb{Q})) - \dim \text{Ker}\,(1-H_{2k}(f) / {}_{\text{Tor}} \otimes \mathbb{Z}_2)) \bmod 2.$$

We call this expression the <u>de Rham invariant of the isometric structure.</u>

If $I(M,f) = 0$ then $w_2(M_f) \cdot w_{4k-1}(M_f)$ vanishes. $H_{2k}(f)/_{Tor}$ means the induced isomorphism on $H_{2k}(M)/_{Tor}$.

For a proof of Lemma 4.4 see Appendix Proposition 8.

Remark 4.5: There is no corresponding statement for the 4k+2-dimensional case.

Remark 4.6: Both dim Ker $(1-H_{2k}(f,\mathbb{Q}))$ mod 2 and dim Ker $(1-H_{2k}(f;\mathbb{Z}_2))$ mod 2 are linear independent bordism invariants of (M,f) but only their difference can be expressed as characteristic number of the mapping torus. In fact it turns out from the main Theorem that the Stiefel-Whitney number $w_2(M_f) \cdot w_{4k-1}(M_f)$ is the only characteristic number of the mapping torus which can be expressed in terms of the isometric structure. It would be interesting to find formulas for other characteristic numbers of M_f.

Problem 4.7: Determine the characteristic numbers of M_f in terms of M and f.

Now we have all the material to describe the image of our invariants. To do this we introduce the following quotient groups of $W_+(\mathbb{Z};\mathbb{Z})$. We consider the element

$$\left[\mathbb{Z}^4, \begin{pmatrix} 1 & & & 0 \\ & 1 & & \\ & & -1 & \\ 0 & & & -1 \end{pmatrix}, \begin{pmatrix} 0 & 1 & & 0 \\ 1 & 0 & & \\ & & 1 & 0 \\ 0 & & 0 & -1 \end{pmatrix}\right].$$

This element has order 2, the vectors (0, 1, 1, 1, 0, 1, 0, 0), (1, 0, 1, -1, 1, 0, 0, 0), (1, 0, 0, 0, -1, 0, 1, -1) and (0, 1, 0, 0, 0, -1, 1, 1) generate an invariant subkernel in 2 times this element.

Definition 4.8: We denote the quotient of $W_+(\mathbb{Z};\mathbb{Z})$ by the subgroup generated by this element and by $[\mathbb{Z},1,1]$ by $\overline{W}_+(\mathbb{Z};\mathbb{Z})$ and the map induced by $I(M,f)$ by $\overline{I}(M,f)$.

The subgroup is isomorphic to $\mathbb{Z} \oplus \mathbb{Z}_2$, the isomorphism is given by $[V,s,g] \longmapsto$ (sign s, dim Ker (1-g) - dim Ker $(1-g \otimes \mathbb{Z}_2)$ mod 2.

Proposition 4.9:

For $k \geqslant 1$ the following homomorphisms are surjective:

$$\theta_{2k-1}: \Delta_{2k-1} \longrightarrow \Omega_{2k-1} \oplus \hat{\Omega}_{2k}, \quad [M,f] \longmapsto ([M],[M_f])$$

$$\theta_{4k-2}: \Delta_{4k-2} \longrightarrow W_-(\mathbb{Z};\mathbb{Z}) \oplus \Omega_{4k-2} \oplus \Omega_{4k-1}, \quad [M,f] \longmapsto (I(M,f),[M],[M_f])$$

$$\theta_{4k} \ : \Delta_{4k} \longrightarrow \overline{W}_+(\mathbb{Z};\mathbb{Z}) \oplus \Omega_{4k} \oplus \Omega_{4k+1}, \quad [M,f] \longmapsto (\overline{I}(M,f),[M],[M_f]).$$

If $\theta_{4k}(M,f) = 0$ then $I(M,f)$ is zero.

Proof: We have shown already the surjectivity of θ_{2k-1}. The surjectivity of θ_{4k-2} and θ_{4k} is obvious for $k = 1$ and follows for $k > 1$ by multiplication with $[P_{2k-2}\mathbb{C}, \mathrm{Id}]$. The last statement follows as $\overline{I}(M,f) = 0$ implies that $I(M,f)$ is contained in the subgroup classified by the signature and the difference of the dimensions of $\mathrm{Ker}(1-f_*)$ with coefficients in $\mathbb{Q}$ and $\mathbb{Z}_2$ mod 2. But $[M] = 0$ and $[M_f] = 0$ and Lemma 4.4 imply that $I(M,f) = 0$.

q.e.d.

Remark 4.10: We can reformulate the surjectivity of θ_{4k} by the following split exact sequence:

$$\Delta_{4k} \longrightarrow W_+(\mathbb{Z};\mathbb{Z}) \oplus \Omega_{4k} \oplus \Omega_{4k+1} \longrightarrow \mathbb{Z} \oplus \mathbb{Z}_2 \longrightarrow 0$$

where the map on the right side is the difference of the signatures in $W_+(\mathbb{Z};\mathbb{Z})$ and Ω_{4k} and of the de Rham invariants in $W_+(\mathbb{Z};\mathbb{Z})$ and Ω_{4k+1}

§ 5 Fibrations over S^1 within their bordism class and the computation of Δ_*

In this chapter we will formulate a result about fibrations over S^1 which as a corollary implies the computation of Δ_*. In § 4 we have mentioned the problem which manifolds within their bordism class are fibrations over S^1. Here we want to investigate the analogous relative problem, too.

The problem is the following. Given a singular manifold $p: N \longrightarrow S^1$ such that $p|_{\partial N}$ is a differentiable fibre bundle. Under which conditions is p bordant rel. boundary to a differentiable fibre bundle $p': N' \longrightarrow S^1$, i.e. there exists a singular manifold with corners and continous map $P: W \longrightarrow S^1$ with $\partial W = N \cup -N' \cup \partial N \times I$ and such that P restricts to the given maps?

To formulate the answer we need a further invariant for diffeomorphisms. Let (M^{4k+2}, f) be a diffeomorphism. On $H_{2k+1}(M, \mathbb{Q})$ we have a symmetric bilinear form:

$$(x,y) \longmapsto (f_* - f_*^{-1})(x) \circ y$$

where $\circ$ denotes the intersection form.

Definition 5.1:

The signature of this bilinear form we denote by $\varphi(M,f)$.

Remark 5.2: φ is no bordism invariant as the following example shows.

Example 5.3: Let $g: S^{2k+1} \times S^{2k+1} \longrightarrow S^{2k+1} \times S^{2k+1}$ be the clutching function of the sphere bundle of the tangent bundle of S^{2k+2}. Then, with respect to the standard basis of $H_{2k+1}(S^{2k+1} \times S^{2k+1})$, g_* has the matrix description $\begin{pmatrix} 1 & 0 \\ 2 & 1 \end{pmatrix}$ and $\circ$ has the matrix description $\begin{pmatrix} 0 & 1 \\ -1 & 0 \end{pmatrix}$. Thus $\varphi(S^{2k+1} \times S^{2k+1}, g) = 1$ but $(S^{2k+1} \times S^{2k+1}, g)$ is null bordant as the sphere bundle bounds the disk bundle.

Now, let N^{4k} be as above such that the boundary of N is a fibre bundle over S^1. We will use φ to compute under certain assumptions the signature of N. Let $f: M \longrightarrow M$ be the classifying diffeomorphism of the fibre bundle ∂N. Let $z \in S^1$ be a regular value of $p: N \longrightarrow S^1$. We cut along $p^{-1}(z) = F$ to obtain a manifold with corners denoted by N_F. Neumann proved the following result.

Proposition 5.4 ([36], Cor. 7.6 + Lemma 8.2): Assume that Ker i_* : $H_{2k+1}(M;\mathbb{Q}) \longrightarrow H_{2k+1}(F;\mathbb{Q})$ is an invariant subkernel of $I(M,f)$.

Then for the signature $\tau(N)$ we have the formula:

$$\tau(N) = \tau(N_F) + \varphi(M,f).$$

In particular if $p : N \longrightarrow S^1$ is a fibre bundle then $\tau(N_F) = 0$ and we have:

$$\tau(N) = \varphi(M,f).$$

Thus $\tau(N) - \varphi(M,f)$ is an obstruction for N to be bordant rel. boundary to a fibre bundle over S^1. Now we can give the complete answer to our problem.

Theorem 5.5: Let $m > 4$ and $p: N^m \longrightarrow S^1$ be a singular manifold whose restriction to ∂N is a differentiable fibre bundle. Let $f : M \longrightarrow M$ be the classifying diffeomorphism of $p|_{\partial N}$.

Then p is bordant rel. boundary to a differentiable fibre bundle over S^1 whose restriction to the boundary is the given one, if and only if

a) m is odd

b) $m \equiv 2(4)$ and $I(M,f) = 0$

c) $m \equiv 0(4)$, $I(M,f) = 0$ and $\tau(N) = \varphi(M,f)$.

Remarks 5.6: 1.) In the proof we don't use the result for $\partial N = \emptyset$. Thus we obtain an independent proof of the result of Neumann and Winkelnkemper (Theorem 4.2).

2.) The case m = 5 was not contained in my Habilitationsschrift. The first proof of it was obtained by P. Melvin [31]. In between I can extend my proof to m = 5 using the stable 5-dimensional h-cobordism Theorem proved by F. Quinn ([37]).

The proof of Theorem 5.5 will be given in the succeeding chapters but first we use it to prove our main result, the computation of Δ_*. Recall that $\hat{\Omega}_m$ denotes the kernel of the signatur and that $\overline{W}_+(\mathbb{Z};\mathbb{Z})$ is the quotient of $W_+(\mathbb{Z};\mathbb{Z})$ as in Definition 4.8.

Theorem 5.7: For $k \geq 1$ the following maps are isomorphisms:

$$\theta_{4k-1}: \Delta_{4k-1} \longrightarrow \Omega_{4k-1} \oplus \hat{\Omega}_{4k}; [M,f] \longmapsto ([M],[M_f])$$

$$\theta_{4k}: \Delta_{4k} \longrightarrow \overline{W}_+(\mathbb{Z};\mathbb{Z}) \oplus \Omega_{4k} \oplus \Omega_{4k+1}, [M,f] \longmapsto (I(M,f),[M],[M_f])$$

$$\theta_{4k+1}: \Delta_{4k+1} \longrightarrow \Omega_{4k+1} \oplus \Omega_{4k+2}, [M,f] \longmapsto ([M],[M_f])$$

$$\theta_{4k+2}: \Delta_{4k+2} \longrightarrow W_-(\mathbb{Z};\mathbb{Z}) \oplus \Omega_{4k+2} \oplus \Omega_{4k+3}, [M,f] \longmapsto (I(M,f),[M],[M_f])$$

Remarks 5.8: 1.) In my Habilitationsschrift [26] I have computed Δ_m only for $m \geq 4$. The computation of Δ_3 was first done by P. Melvin [31]. As mentioned in Remark 5.6 my proof extends now to this case, too.
2.) Analogously to Remark 4.10 we can replace the statement about θ_{4k} by the split short exact sequence:

$$0 \to \Delta_{4k} \longrightarrow W_+(\mathbb{Z};\mathbb{Z}) \oplus \Omega_{4k} \oplus \Omega_{4k+1} \longrightarrow \mathbb{Z} \oplus \mathbb{Z}_2 \longrightarrow 0.$$

We will see in § 9 that this is a formulation which can be extended to the case of manifolds with an additional structure (Theorem 9.9).

As a consequence of Theorem 2 we can determine odd dimensional diffeomorphism classes by numerical invariants and we can do the same for even-dimensional diffeomorphisms up to torsion.

Corollary 5.9: Two odd-dimensional diffeomorphisms are bordant if and only if all Stiefel-Whitney numbers of the underlying manifold and all characteristic numbers of the mapping torus are the same.

For two even-dimensional diffeomorphisms (M_1,f_1) and (M_2,f_2) of dim > 2 the following holds: $4 \cdot (M_1,f_1)$ and $4 \cdot (M_2,f_2)$ are bordant if and only if for all $n \in \mathbb{Z}$ the equivariant signatures sign (M_1,f_1^n) = sign (M_2,f_2^n) and all Pontrjagin numbers of M_1 and M_2 are the same.

We will discuss special cases and applications of the Theorems 5.5 and 5.7 in later chapters. Here we first want to show how Theorem 5.7 follows from

Theorem 5.5 and then we want to indicate the idea of the proof of Theorem 5.5.

Reduction of Theorem 5.7 to Theorem 5.5:

By proposition 4.9 θ_* is surjective so we only have to show the injectivity. Consider a diffeomorphism (M^n,f) with $\theta_n(M,f) = 0$. The projection $p : M_f \longrightarrow S^1$ represents an element of $\Omega_{n+1}(S^1)$, the bordism group of singular manifolds in S^1. This group is isomorphic to $\Omega_{n+1} \oplus \Omega_n$ ([13]), the isomorphism is given by forgetting p and the fibre over a regular point. Thus as $[M] = 0$ and $[M_f] = 0$ there exists a n+2-dimensional singular manifold $p : N \longrightarrow S^1$ with boundary $p : M_f \longrightarrow S^1$. Now for n odd or $n = 0(4)$ it follows from Theorem 5.5 that we can replace $p : N \longrightarrow S^1$ by a differentiable fibre bundle and so (M,f) bounds.

For $n = 2(4)$ we would be finished if $\tau(N) = \varphi(M,f)$. To achieve this it is enough to construct a singular manifold $q : V \longrightarrow S^1$ with fibred boundary and classifying diffeomorphism $g: T \longrightarrow T$ such that $\tau(V) - \varphi(T,g)=1$ and (T,g) is null-bordant. For then we change (M,f) and $p: N \longrightarrow S^1$ by adding an appropriate multiple of (T,g) and $q: V \longrightarrow S^1$.

An example for such a V is given by $P_k\mathbb{C}$, $2k = n+2$, and q the constant map. Here the boundary and thus T is empty and we have $\tau(V) - \varphi(T,g)= \tau(V)=1$.

q.e.d.

We want to end this chapter with a short discussion of the idea of the proof of Theorem 5.5: Consider a singular manifold $p: N \longrightarrow S^1$ such that

$p|_{\partial N}$ is a differentiable fibre bundle with classifying diffeomorphism $f: M \longrightarrow M$. Then we introduce the manifold N_F as follows. Let $z \in S^1$ be a regular value of p. As we consider the problem only up to bordism we can assume that $p^{-1}(z)$ is non empty. For if it would be empty we can add the singular manifold $(S^1 \times F, p_1)$ where p_1 is the projection onto S^1 and F is any closed zero-bordant (m-1)-dimensional manifold. If $p^{-1}(z)$ is non empty then $F = p^{-1}(z)$ is a two-sided manifold in N. We cut N along T to obtain a manifold denoted by N_F. If we straighten the angle of N_F we obtain a smooth manifold whose boundary is equal to the twisted double $F \cup_f - F$ where $f: M \longrightarrow M$ is the classifying diffeomorphism of the fibre bundle $p|_{\partial N}$. We call such a manifold N_F an <u>admissible manifold</u> for the diffeomorphism (M,f).

<u>Definition 5.10:</u> Let (M,f) be a diffeomorphism. We call a compact oriented differentiable manifold <u>admissible for (M,f)</u> if $\partial N_F = F \cup_f - F$.

<u>Remark 5.11:</u> We have seen above that a singular manifold $p: N \longrightarrow S^1$ with $\partial N = M_f$ leeds to an admissible manifold for (M,f). On the other hand given an admissible manifold N_F we can construct $p: N \longrightarrow S^1$ with $\partial N = M_f$ which by the construction above leeds back to N_F. We do this as follows.

We first introduce the submanifold F_0 and F_1. We do this by removing a bi-collar of $M = \partial F$ in ∂N_F from ∂N_F, the resulting compliment consists of two copies of F denoted by F_0 and F_1. If we identify F_0 and F_1 and remove the corners ([13], I.3) we obtain a manifold denoted by N with $\partial N = M_f$. It is not difficult to show that the projection of ∂N to S^1 extends to a continous map $p: N \longrightarrow S^1$ such that $1 \in S^1$ is a regular value and $p^{-1}(1)=F$. This property classifies p up to homotopy. Thus p is unique up to homotopy.

Proposition 5.12: (compare [11] , 2.3) Let N_F be an admissible manifold for a diffeomorphism (M,f), dim $N_F > 5$ such that the components of F are 1-connected. Then f can be extended to a diffeomorphism of F if and only if the inclusions of F_0 and F_1 into N_F are homotopy equivalences.

In particular this is fulfilled if F and N_F are 1-connected and $\pi_k(N_F,F_i) = \{0\}$ for all k and i = 0, 1.

The same statement is true for dim $N_F = 5$ if we stabilize N_F by adding an appropriate number of $S^2 \times S^2 \times I$.

Proof: If we introduce corners along the boundary of F_0 and F_1 we see that N_F is a relative h-cobordism between F_0 and F_1. Thus the proposition follows for dim $N_F > 5$ from the relative h-cobordism Theorem [41] . The case dim $N_F = 5$ follows from the stable h-cobordism Theorem [37] .

q.e.d.

Remark 5.13: If N_F and F are 1-connected and dim $N_F = n$ then the conditions of proposition 5.12 are equivalent to : $H_k(N_F,F_i) = \{0\}$ for $k \leq n/2$ and i = 0, 1.

This follows from the Hurewicz isomorphism and the Poincaré duality $H_k(N_F,F_0) \cong H^{n-k}(N_F,F_1)$ [39] .

We now sketch the proof of Theorem 5.5 as follows. Given $p : N \longrightarrow S^1$ satisfying the conditions of Theorem 5.5 we consider an admissible manifold N_F for (M,f), the classifying diffeomorphism of $p|_{\partial N}$. Then we change N_F by certain modifications similar to addition and subtraction of handles to obtain a new admissible manifold $N'_{F'}$ which fulfils the conditions of Proposition 5.12. We do this so that the resulting fibre bundle $p' : N' \longrightarrow S^1$ is bordant rel. boundary to p.

§ 6 Addition and subtraction of handles

For a singular manifold $p : N \longrightarrow S^1$ whose boundary is a differentiable fibre bundle over S^1 classified by a diffeomorphism (M,f) we have introduced an admissible manifold N_F for (M,f). In this chapter we define some modifications of N_F and, consequentely, of N and discuss how they change the homology- and homotopy groups. The first modification is surgery on N_F. In N this corresponds to surgery of N which doesn't change the fibre. It is obvious that the resulting manifold N' is bordant to N rel. boundary such that p can be extended to this bordism. We need two other types of modifications: addition and subtraction of handles.

Before we define these modifications we introduce a bordism relation for admissible manifolds which has the property that if N_F and $N'_{F'}$ are bordant then the corresponding singular manifolds $p : N \longrightarrow S^1$ and $p' : N' \longrightarrow S^1$ as constructed in Remark 5.11 are bordant rel. boundary.

Definition 6.1: Let (M,f) be a fixed diffeomorphism. Two admissible manifolds N_F and $N'_{F'}$ for (M,f) are called bordant if there exist manifolds V and W such that V is a bordism rel. boundary between F and F', i.e. $\partial V = F \cup M \times I \cup F'$, and $W = N_F \cup (V \underset{f\times Id}{\cup} -V) \cup N'_{F'}$, where $V \underset{f\times Id}{\cup} -V$ is the manifold obtained by identifying $M \times I$ in V and $-V$ by $f\times Id$ and straightening the angle.

If (W,V) is such a bordism between N_F and $N'_{F'}$ then as in Remark 5.11 we can identify V and $-V$ in ∂W to obtain a manifold $\tilde{W}$ and a continuous map $p : \tilde{W} \longrightarrow S^1$. (W,p) is a bordism between (N,p) and (N',p'). So we obtain

Lemma 6.2: If two admissible manifolds N_F and $N'_{F'}$ for (M,f) are bordant then the corresponding singular manifolds as constructed in Remark 5.11 are bordant rel. boundary.

a) Addition of handles. Let N_F be a n-dimensional admissible manifold for a diffeomorphism (M,f). Let $l : S^k x D^{n-k} \longrightarrow \mathring{F}$ be a differentiable embedding. We denote the corresponding embedding into F_o and F_1 by l_o and l_1 resp. Now we glue two handles $D^{k+1} x D^{n-k-1}$ to N_F with l_o and l_1 and obtain (after straightening the angle) an admissible manifold for (M,f) denoted by $N'_{F'} := N_F \cup_{l_o} D^{k+1} \times D^{n-k-1} \cup_{l_1} D^{k+1} \times D^{n-k-1}$. F' is obtained from F by surgery with l. We say that $N'_{F'}$ is obtained from N_F by addition of handles with l.

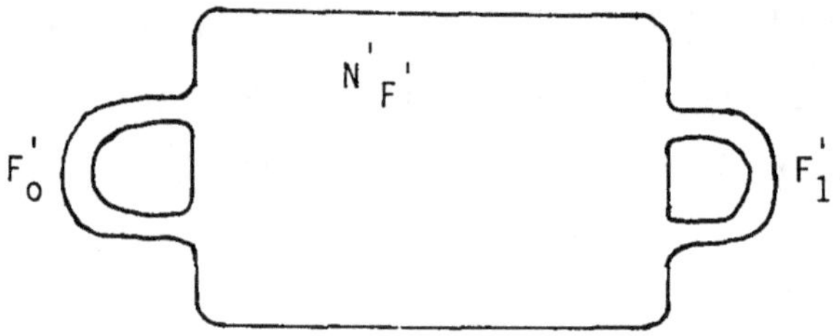

It is obvious that N_F and $N'_{F'}$ are bordant. Thus we obtain

Proposition 6.3: If $N'_{F'}$ is obtained from N_F by addition of handles as above then N_F and $N'_{F'}$ are bordant and as a consequence the singular manifolds $p : N \longrightarrow S^1$ and $p' : N' \longrightarrow S^1$ are bordant rel. boundary.

We want to replace a singular manifold $p : N \longrightarrow S^1$ with fibred boundary within its bordism class rel. boundary by a singular manifold $p' : N' \longrightarrow S^1$ such that $N'_{F'}$ and F' are 1-connected. This can be obtained by surgery on

N_F and addition of handles. We formulate this in the following Lemma.

Lemma 6.4: A singular manifold $p : N \longrightarrow S^1$ with fibred boundary is bordant rel. boundary to a singular manifold $p' : N' \longrightarrow S^1$ such that $N'_{F'}$ and F' are 1-connected.

Proof: It is well known that we can replace F by a sequence of surgeries in the interior of F by a 1-connected manifold. Thus by a sequence of additions of handles we can replace N_F by $N'_{F'}$ such that F' is 1-connected. By the same argument we can replace $N'_{F'}$ by a sequence of surgeries in the interior of $N'_{F'}$ by a 1-connected manifold.

q.e.d.

Remark: In the following we always assume for $p : N \longrightarrow S^1$ that N_F and F are 1-connected.

We now describe how the homotopy groups of (N_F,F_i) and of F are altered if we add handles as above.

Lemma 6.5: Let N_F be a n-dimensional admissible manifold with N_F and F 1-connected. Furthermore we suppose that there exists a k with $1 < k \leq (n-1)/2$ such that $\pi_r(N_F,F_i) = \{0\}$ for $1 < r < k$ and $i = 0$. Let $l: S^{k-1} x D^{n-k} \longrightarrow \mathring{F}$ be a differentiable embedding and $\alpha \in \pi_{k-1}(F)$ the element represented by $l(S^{k-1} x \{0\})$.

The manifold $N'_F{}'$ obtained from N_F by addition of handles with 1 has the following properties:

1.) $N'_F{}'$ and F' are 1-connected

2.) $\pi_r(N'_F{}', F'_i) = \{0\}$ for $1 < r < k$ and $i = 0, 1$
1.) and 2.) also hold if n is even and $k = n/2 - 1$.

3.) $\pi_{k-1}(F') \cong \pi_{k-1}(F) /_{(\alpha)}$ and $\pi_r(F') = \pi_r(F)$ for $r < k-1$
where (α) denoted the subgroup generated by α.

The proof is standard and can be found for instance in [26].

We denote the boundary operators $\pi_k(N_F, F_i) \longrightarrow \pi_{k-1}(F)$ by d^i. Let N_F be as in Lemma 6.5 and $k \leq (n-1)/_2$. Then every element of image d^i can be represented by an embedded sphere with trivial normal bundle and we can apply Lemma 6.5 to kill this element. The reason is that the sphere bounds an embedded disk in N_F.

Lemma 6.6: Let N_F be an admissible manifold as in Lemma 6.5. Then by a sequence of handle additions we can replace N_F by an admissible manifold $\bar{N}_{\bar{F}}$ with the same properties as N_F and such that in addition the boundary operators $\partial^i : \pi_k (\bar{N}_{\bar{F}}, \bar{F}_i) \longrightarrow \pi_{k-1}(\bar{F})$ are zero for $k \leq (n-1)/_2$.

b) Subtraction of handles. Let $k \leq n/_2$ and $s : S^k \longrightarrow \mathring{N}_F$ be a differentiable embedding into the n-dimensional admissible manifold N_F. If $k = n/_2$ we suppose that the self intersection number of this embedding is zero. Then we choose disjoint embeddings s_0 and s_1 isotopic to s. Furthermore we choose two disjoint embedded paths w_0 and w_1 joining F_0 with $s_0(S^k)$ and F_1 with

$s_1(S^k)$. We assume that these paths meet F_0 and $s_0(S^k)$ and F_1 and $s_1(S^k)$ transversally. Let U_0 and U_1 be disjoint differentiable regular neighborhoods of $s_0(S^k) \cup w_0(I)$ and of $s_1(S^k) \cup w_1(I)$. U_i is a n-dimensional manifold with boundary which is homotopy equivalent to $s_i(S^k)$.

Now we remove $\mathring{U}_0$ and $\mathring{U}_1$ from N_F and straighten the angle. The result is a manifold with boundary $F_0' \underset{f}{\cup} F_1'$ where $F_i' = F_i \# S\nu(s_i(S^k))$, $S\nu(s_i(S^k))$ the sphere bundle of the normal disk bundle of $s_i(S^k)$.

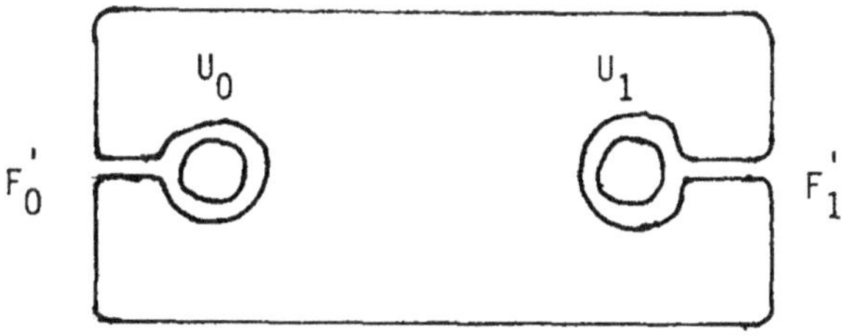

Now we want to show that this manifold is again admissible. For this we have to show that F_0' and $-F_1'$ are diffeomorphic under a diffeomorphism which is the identity on the boundary $\partial F_0' = \partial F_1' = M$. As s_0 and s_1 are isotopic their normal sphere bundles are equal. Furthermore these bundles admit a section and thus they have an orientation reversing diffeomorphism and $S\nu(s_0(S^k)) = -S\nu(s_1(S^k))$. As $F_0 = -F_1$ it follows that F_0' and $-F_1'$ are diffeomorphic under a diffeomorphism which is the identity on the boundary.

Thus the manifold constructed above is again admissible and we denote it by $N'_{F'}$. We say that $N'_{F'}$ is obtained from N_F by subtraction of handles with s_0 and s_1.

Remark 6.7: This subtraction of handles is a special case of a more general construction which is the inverse process of the addition of handles and makes the notation subtraction of handles more clear (compare [49]). Consider disjoint embeddings s_0 and s_1 of (D^k,S^{k-1}) into (N_F,F_i) meeting F_i transversally and remove disjoint open tubular neighborhoods from N_F.

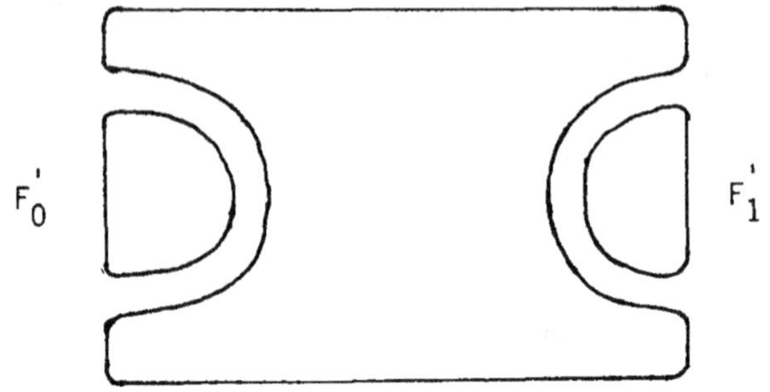

The boundary of the resulting manifolds is equal to $F_0' \underset{f}{\cup} F_1'$. But it is not clear that $F_0' = -F_1'$. Thus in general the resulting manifold is not again admissible.

We have an analogous statement as Proposition 6.3.

Proposition 6.8: If $N'_{F'}$ is obtained from N_F by subtraction of handles then N_F and $N'_{F'}$ are bordant and so the singular manifolds $p : N \longrightarrow S^1$ and $p' : N' \longrightarrow S^1$ are bordant rel. boundary.

In the following we want to use subtraction of handles to kill elements in $\pi_k(N_F,F_i)$. By our subtraction of handles we can do this only for elements which can be represented by spheres in N_F. To achieve this we first have introduced addition of handles which allows to replace an admissible manifold by one for which every element of $\pi_k(N_F,F_i)$ can be represented by a sphere in N_F (Lemma 6.6). If we use such a sphere to make subtraction of handles it is obvious that the element in $\pi_k(N_F,F_i)$ represented by the sphere is killed as it is representable then by a sphere in F_i and if $k \leq n/_2-1$, n the dimension of N_F, the homotopy group doesn't increase. More precisely we formulate this in the following Lemma.

Lemma 6.9: Let N_F be a n-dimensional admissible manifold such that N_F and F are 1-connected and for a $k < n/_2$ the homotopy groups $\pi_r(N_F,F_i)$ vanish for $1 < r < k$ and $i = 0,1$. Let s_0 and $s_1 : S^k \longrightarrow \mathring{N}_F$ be disjoint isotopic differentiable embeddings and α the element in $\pi_k(N_F)$ represented by these embeddings. We denote the elements in $\pi_k(N_F,F_i)$ represented by α by α_i. Let $N'_F{}'$ be obtained by subtraction of handles with s_0 and s_1.

Then we have

1.) $N'_F{}'$ and F' are 1-connected

2.) $\pi_r(N'_F{}',F'_i) = \{0\}$ for $1 < r < k$ and $i = 0, 1$.

If $k \leq n/_2-1$ we have

$\pi_k(N'_F{}',F'_i) \cong \pi_k(N_F, F_i) /_{(\alpha_i)}$

where (α_i) denotes the subgroup generated by α_i.

3.) If the boundary operators $d^i : \pi_k(N_F,F_i) \longrightarrow \pi_{k-1}(F)$ vanish then the boundary operators $d'^i: \pi_k(N'_F{}',F'_i) \longrightarrow \pi_{k-1}(F'_i)$ are also zero.

4.) $\pi_r(N'_F{}') \cong \pi_r(N_F)$ for $r \leq k$.

The proof is again standard and can be found in ([24] , Lemma 5).

Now, we have all the machinery necessary to kill the relative homotopy groups $\pi_r(N_F,F_i)$ below the middle dimension. We will see that for N_F odd-dimensional we also can kill the middle dimensional homotopy groups but in the even-dimensional case we need stronger assumptions to (N_F,F_i) which allow to translate the assumptions of Theorem 5.5 into conditions to (N_F,F_i) under which we can kill the middle dimensional homotopy groups too. Thus we give separate formulations for dim N_F odd and even.

Proposition 6.10: Let N_F be a 2k+1-dimensioned admissible manifold ($k>1$). By a sequence of surgeries, additions and subtractions of handles we can replace N_F by an admissible manifold $N'_{F'}$ with the following properties:

1.) $N'_{F'}$ and F' are 1-connected

2.) $\pi_r(N'_{F'}, F'_i) = \{0\}$ for $r < k$ and $i = 0,1$

3.) $d^i : \pi_k(N'_{F'}, F'_i) \longrightarrow \pi_{k-1}(F'_i)$ is zero

Proof: By Lemma 6.4 we can assume that N_F and F are 1-connected. By induction we can assume that $\pi_r(N_F, F_i) = \{0\}$ for $r < s$. If $s \leq k$ we can kill d^s by a sequence of additions of handles (Lemma 6.6). But then we can, for $s < k$, apply Lemma 6.9 to kill $\pi_s(N'_{F'}, F'_i)$.

q.e.d.

For the next Proposition we need the definition of the strong self linking number ϱ in the following situation. Let F be $(2k-1)$-dimansional and k odd. Given $x \in \pi_{k-1}(F)$ such that the Hurewicz $H(x)$ is torsion in $H_{k-1}(F)$ and $\nu_*(x) = 0$ in $\pi_{k-1}(BSO)$. Then we can represent x by an embedding $S^{k-1} \times D^k \longrightarrow F$ denoted by U. We consider the following diagram of exact sequences:

$$\begin{array}{ccccccccc} & & & & Z & & 1 & & \\ & & & & \downarrow & & \downarrow & & \\ & & & & & & e_1 & & \\ 0 & \longrightarrow & Z & \longrightarrow & H_{k-1}(F - \mathring{U}) & \longrightarrow & H_{k-1}(F) & \longrightarrow & 0 \\ & & 1 & \longmapsto & e_2 & & & & \\ & & & & \downarrow & & & & \\ & & & & H_{k-1}(F') & & & & \\ & & & & \downarrow & & & & \\ & & & & 0 & & & & \end{array}$$

F' is the result of surgery with this embedding and e_1 and e_2 are represented by $S^{k-1} \times \{*\}$ and $\{*\} \times S^{k-1}$. If H(x) has order r then there exists a $s \in Z$ such that in $H_{k-1}(F - \mathring{U})$:

$r e_1 + s e_2 = 0.$

If $k \neq 2,4,8$ the number $r/s \in Q/2Z$ depends only of x and is denoted by $\varrho(x)$ (compare [10], p.105). For k = 2,4,8 we define $\varrho(x) = L(x,x) \in Q/Z$ as the ordinary self linking number r/s mod 1.

Proposition 6.11: Let N_F be a 2k - dimensional admissible manifold ($k > 2$). By a sequence of surgeries and addition of handles we can replace N_F by $N'_{F'}$ with the following properties:

The normal Gauß map $\nu : F' \longrightarrow BSO$ is a (k-1) - equivalence and $\nu : N'_{F'} \longrightarrow BSO$ is a k - equivalence, and for k even, $k \neq 2,4,8$ there exists a $x \in \pi_{k-1}(F')$ such that the strong self linking number $\varrho(x) \in Q/2Z$ is 1, H(x) = 0 and $\nu_*(x) = 0$.

Proof: The last condition is not touched by surgeries on F below the dimension k-1. So, we can achieve this condition first by making the connected sum of F with a zero bordant closed manifold X which has this property.

For the construction of X we consider the diagonal $S^{k-1} \subset S^{k-1} \times S^{k-1}$. The normal bundle of this embedding is the tangent bundle of S^{k-1}. Let $E \longrightarrow Y$ ba a (k-1) - dimensional oriented vector bundle over an oriented k - dimensional compact manifold Y with $\partial Y = S^{k-1}$ and $E|_{\partial Y}$ the tangent bundle of S^{k-1}. If we glue E to $S^{k-1} \times D^k$ and double the resulting manifold with boundary we obtain a manifold X. $S^{k-1} \times \{0\} \subset S^{k-1} \times D^k \subset X$ represents zero in $H_{k-1}(X)$ and it is obvious that $\varrho(S^{k-1} \times \{0\}) = 1$.

The existence of such a (k-1) - dimensional oriented vector bundle E over

Y is equivalent to the existence of a k - dimensional oriented vector bundle with Euler class 1 over $Y \cup D^k$. We can obtain E from such a bundle by constructing a section with only one isolated singularity in $0 \in D^k$, removing $\mathring{D}^k$ and taking the orthogonal complement of this section. A k - dimensional vector bundle over a closed oriented k - manifold with this property is given by the product of k/2 factors of the Hopf bundle $H \longrightarrow S^2$ which has Euler class 1.

The idea to achieve the first condition is the following. By a finite sequence of appropriate additions of handles we can obtain a manifold $N'_{F'}$ s.t. $\nu_*: \pi_r(F') \longrightarrow \pi_r(BSO)$ is surjective for $r \leq k-1$. We do this by making a handle addition for each generator of $\pi_*(BSO), * < k$, with an appropriately twisted embedding of a zero homotopic sphere in F. Furthermore we can do surgery on $N'_{F'}$ to assume that in addition $\pi_k(N'_{F'}) \longrightarrow \pi_k(BSO)$ is surjective. By a general position argument this surjectivity is not affected by further surgeries and addition of handles below the middle dimension. The elements in the kernel of $\pi_r(F') \longrightarrow \pi_r(BSO)$ for $r < k-1$ or of $\pi_r(N'_{F'}) \longrightarrow \pi_r(BSO)$ for $r < k$ can be represented by spheres with trivial normal bundle and thus inductively be killed by a sequence of handle additions and of surgeries.

q.e.d.

§ 7 Proof of Theorem 5.5 in the odd-dimensional case

By Remark 5.13 the proof of Theorem 5.5 is completed if we can show that every n-dimensional admissible manifold N_F with $I(M,f) = 0$ and $\tau(N) = \varphi(M,f)$ can be replaced by a sequence of surgeries and additions and subtractions of handles by an admissible manifold $N'_{F'}$ such that $N'_{F'}$ and F' are 1-connected and $H_k(N_F,F_i) = \{0\}$ for $k \leq n/_2$ and $i = 0,1$.

For n odd, $n > 5$, we have shown this in [24] . For completeness we indicate here the most important steps of the proof. Let N_F be a 2k+1-dimensional admissible manifold. By Proposition 6.10 we can assume the following properties:

N_F and F are 1-connected, $\pi_r(N_F,F_i) = \{0\}$ for $r < k$ and d^i :
$\pi_k(N_F,F_i) \longrightarrow \pi_{k-1}(F)$ is zero.

The last condition implies that every element of $\pi_k(N_F,F_i)$ can be represented by an embedding $S^k \longrightarrow \mathring{N}_F$. We wish to kill $\pi_k(N_F,F_i)$. It is not difficult to show that $\pi_k(N_F,F_0)$ and $\pi_k(N_F,F_1)$ are isomorphic ([24] , Lemma 9) thus it is sufficient to kill $\pi_k(N_F,F_0)$.

Let $s_0,s_1 : S^k \longrightarrow \mathring{N}_F$ be two disjoint isotopic embeddings. By subtraction of handles with s_i we obtain the admissible manifold $N'_{F'}$. We denote the element in $\pi_k(N_F,F_0)$ represented by $s_0(S^k)$ by α and the element in $\pi_k(N'_{F'},F'_0)$ represented by a fibre of the normal sphere bundle of $s_1(S^k)$ by β .

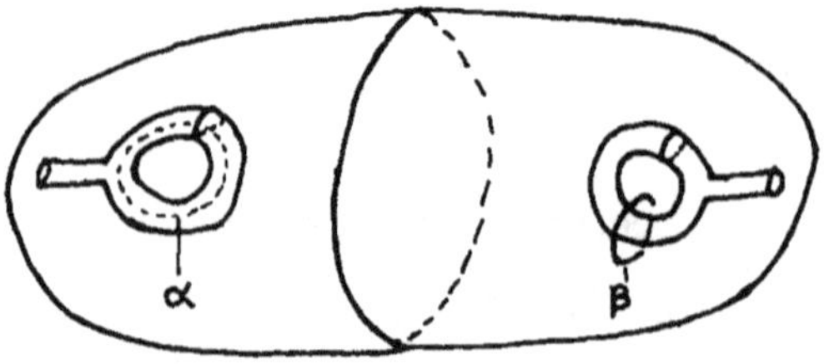

Then $\pi_k(N_F,F_0)/_{(\alpha)} \cong \pi_k(N'_F{}',F'_0)/_{(\beta)}$ ([24] , p. 359).

The behaviour of β can be controlled by the element represented by $s_0(S^k)$ in $\pi_k(N_F,F_1)$ denoted by γ. For if $\gamma \in \pi_k(N_F,F_1)$ is primitive, i.e. a generator of an infinite direct summand, then β is zero and

$$\pi_k(N_F,F_0)/_{(\alpha)} \cong \pi_k(N'_F{}',F'_0) \qquad ([24] , p.359).$$

This fact allows us to replace $\pi_k(N_F,F_0)$ by a torsion group. For if $\pi_k(N_F,F_0)$ is infinite there exist disjoint isotopic embeddings s_0 and s_1 such that α has infinite order and γ is primitive or α is primitive and γ has infinite order ([24] , p.359). By subtractions of handles with s_i the rank of $\pi_k(N_F,F_i)$ decreases and can be killed after a finite step of those subtractions of handles.

Now we suppose that $\pi_k(N_F,F_0)$ is a torsion group. Let $\alpha \in \pi_k(N_F,F_0)$ be a non-trivial element represented by $s_0\colon S^k \longrightarrow N_F$. We have shown in ([24] , p.359) that we can find an embedding $s_1\colon S^k \longrightarrow N_F$ isotopic to s_0 such that β in $\pi_k(N'_F{}',F'_0)$ has infinite order or smaller order than α, where $N'_F{}'$ is obtained from N_F by subtraction of handles with s_0 and s_1. If β has finite order this implies that $\pi_k(N'_F{}',F'_0)$ is smaller than $\pi_k(N_F,F_0)$. If β has infinite order, $\beta = m \cdot x$, where $x \in \pi_k(N'_F{}',F'_0)$ is primitive and $m \in \mathbb{N}$. Then the torsion subgroup of $\pi_k(N'_F{}',F'_0)$ has order equal to the order of $\pi_k(N_F,F_0)$ divided by the order of α times m. Now we can find an element y in $\pi_k(N'_F{}')$ such that for the elements α' and γ' represented by y in $\pi_k(N'_F{}',F'_0)$ and $\pi_k(N'_F{}',F'_1)$ we have: γ' is primitive and $\alpha' = n \cdot x$ + torsion with $0 < n \leq m$. To find this we take an arbitrary y with γ' primitive and modify it by addition of an appropriate multiple of β considered as an element of $\pi_k(N'_F{}')$. For a y with the properties above we choose disjoint isotopic embeddings $S^k \longrightarrow N'_F{}'$ and do the subtraction of handles. The result is an admissible manifold $\bar{N}_{\bar{F}}$ with $\pi_k(\bar{N}_{\bar{F}},\bar{F}_0) \cong \pi_k(N'_F{}',F'_0)/_{(\alpha')}$. Thus $\pi_k(\bar{N}_{\bar{F}},\bar{F}_0)$ is torsion and its order is equal to the order of the torsion subgroup of $\pi_k(N'_F{}',F'_0)$ times m. But this is equal to the order of $\pi_k(N_F,F_0)$ times n divided by the order of α times m and this is smaller than the order of $\pi_k(N_F,F_0)$. Hence in the case where β has infinite order we can also make the order smaller by two subtractions of handles.

So by a finite sequence of subtraction of handles we can kill $\pi_k(N_F,F_0) = H_k(N_F,F_0)$ and this completes the proof of Theorem 5.5 for n odd.

§ 8 Proof of Theorem 5.5 in the even-dimensional case

We first sketch the proof. Consider a $2\cdot k$-dimensional admissible manifold N_F satisfying the conditions of $N'_F{}'$ in Proposition 6.11. We want to kill $\pi_k(N_F,F_i)$. In contrast to the odd-dimensional case there are several additional problems. First we can't, in general, satisfy the condition that the boundary operator $d^i : \pi_k(N_F,F_i) \longrightarrow \pi_{k-1}(F)$ be zero. Hence we can't represent every element in $\pi_k(N_F,F_i)$ by a sphere embedded in N_F to be used in our subtraction of handles. A further difficulty is that even if we can represent an element in $\pi_k(N_F,F_i)$ by an embedded sphere in N_F we may use it to perform subtraction of handles if and only if the self intersection number of the sphere vanishes. Thus there are several difficulties killing $\pi_k(N_F,F_i)$ and in the following we have to build up a connection between these difficulties and the isometric structure.

First we state sufficient conditions for $\pi_k(N_F,F_i)$ which allow it to be killed by subtractions of handles. We will see from these conditions that it is sufficient to represent half of $\pi_k(N_F,F_i)$ by embedded spheres in N_F with trivial self intersection number.

Proposition 8.1: Let N_F be a $2\cdot k$-dimensional admissible manifold such that N_F and F are 1-connected and $\pi_r(N_F,F_i) = \{0\}$ for $1 < r < k$ and $i = 0,1$. We assume that there exist q disjoint embeddings $s_1,\ldots,s_q\colon S^k \longrightarrow N_F$ with vanishing self intersection number, $q = \operatorname{rank} \pi_k(N_F,F_i)/2$. Further we assume that the homotopy classes represented by $s_i(S^k)$ can be extended to a basis of both $\pi_k(N_F,F_0)$ and $\pi_k(N_F,F_1)$. Then we can kill $\pi_k(N_F,F_i)$ by a sequence of subtraction of handles with $s_1,\ldots,s_q$.

Remark 8.2: Under these assumptions $\pi_k(N_F,F_i) \cong H_k(N_F,F_i)$. Poincaré-duality and the universal coefficient Theorem for cohomology imply that $\pi_k(N_F,F_0) \cong \pi_k(N_F,F_1)$ and the groups are torsion free.

Proof: Let s_1^0 and $s_1^1 : S^k \longrightarrow N_F$ be disjoint embeddings isotopic to s_1. Let U_0 and U_1 be regular neighborhoods (see § 6b) and assume that $U_i \cap s_j(S^k) = \emptyset$ for $j > 1$. Let $N'_F{}'$ be obtained from N_F by subtraction of handles with U_0 and U_1. Next we compute the homotopy groups of $N'_F{}'$. From the van Kampen's Theorem it follows that $N'_F{}'$ and F' are again 1-connected.

For $r < k-1$ we have the following diagram of exact sequences with $\mathbb{Z}$-coefficients.

$$\begin{array}{ccccccc}
 & & 0 & & & & \\
 & & \downarrow & & & & \\
0 & \longrightarrow & H_r(N_F - U_1,F_0) & \longrightarrow & H_r(N_F,F_0) & \longrightarrow & 0 \\
 & & \downarrow & & & & \\
 & & H_r(N_F - U_1,F_0 \cup U_0) \cong H_r(N'_F{}',F'_0) & & & & \\
 & & \downarrow & & & & \\
 & & 0 & & & &
\end{array}$$

This implies for $r < k - 1$: $H_r(N'_F{}',F'_0) \cong H_r(N_F,F_0) = \{0\}$ and analogously $H_r(N'_F{}',F'_1) = \{0\}$.

To investigate the cases $r = k - 1$ and $r = k$ we consider the same diagram:

$$
\begin{array}{ccccccc}
 & & H_k(F_0 \cup U_0, F_0) \cong \mathbb{Z} \ni 1 & \longmapsto & [s_1^0(S^k)] & & \\
 & & \downarrow & & & & \\
0 & \longrightarrow & H_k(N_F - U_1, F_0) & \longrightarrow & H_k(N_F, F_0) & \longrightarrow & H_k(N_F, N_F - U_1) \longrightarrow \\
 & & & & x & \longmapsto & x \circ [s_1^1(S^k)] \\
 & & \downarrow & & & & \\
 & & H_k(N_F - U_1, F_0 \cup U_0) \cong H_k(N'_{F'}, F'_0) & & & & \\
 & & \downarrow & & & & \\
 & & 0 & & & &
\end{array}
$$

$$
\begin{array}{ccccccc}
 & & 0 & & & & \\
 & & \downarrow & & & & \\
\longrightarrow & & H_{k-1}(N_F - U_1, F_0) & \longrightarrow & H_{k-1}(N_F, F_0) & \longrightarrow & 0 \\
 & & \downarrow & & & & \\
 & & H_{k-1}(N_F - U_1, F_0 \cup U_0) \cong H_{k-1}(N'_{F'}, F'_0) & & & & \\
 & & \downarrow & & & & \\
 & & 0 & & & &
\end{array}
$$

The map $H_k(N_F, F_0) \longrightarrow H_k(N_F, N_F - U_1) \cong \mathbb{Z}$ is given by $x \longmapsto x \circ [s_1^1(S^k)]$ which can be verified as follows: $H_k(N_F, N_F - U_1)$ is generated by a fibre D^k of the normal disk bundle to $s_1^1(S^k)$ which has intersection number 1 with $[s_1^1(S^k)]$. If we denote the image of $x \in H_k(N_F, F_0)$ in $H_k(N_F, N_F - U_1)$ by x' we have $x \circ [s_1^1(S^k)] = x' \circ [s_1^1(S^k)]$. With $x' = n \cdot [D^k]$ we obtain $n = x \circ [s_1^1(S^k)]$.

Since $[s_1^1(S^k)] \in H_k(N_F, F_1)$ is primitive, Poincaré duality implies that there exists $y \in H_k(N_F, F_0)$ with intersection number $y \circ [s_1^1(S^k)] = 1$.

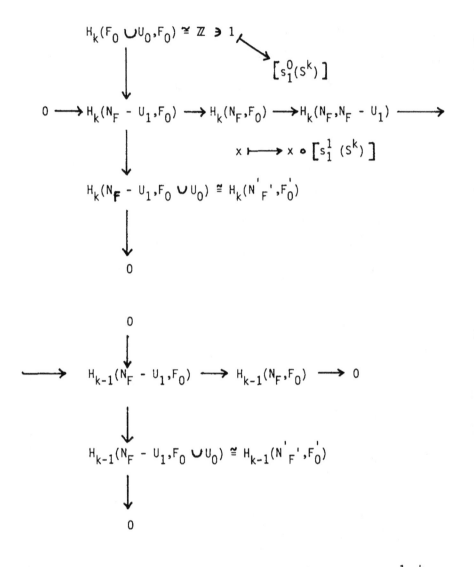

Thus the map $H_k(N_F,F_0) \longrightarrow H_k(N_F,N_F - U_1)$ is surjective and therefore $H_{k-1}(N'_F{}',F'_0) \cong H_{k-1}(N_F,F_0) = \{0\}$.

Furthermore, we may choose y so that $y \circ \left[s_j(S^k) \right] = 0$ for $j > 1$. The diagram implies that

$$H_k(N'_F{}',F'_0) \cong H_k(N_F,F_0)/_{(y,\ [s_1^0(S^k)])}.$$

An analogous statement holds for $H_k(N'_F{}',F'_1)$. From the intersection conditions we see that the embeddings $s_2,\dots,s_q : S^k \longrightarrow N_F$ induce embeddings $s'_2,\dots,s'_q : S^k \longrightarrow N'_F{}'$ with the same properties as for N_F. Thus we can repeat our argument and kill $H_k(N_F,F_i) = \pi_k(N_F,F_0)$ inductively.

q.e.d.

Now we want to show a connection between the vanishing of the isometric structure and the conditions of Proposition 8.1. We can do this only if, in addition to the properties of Proposition 6.11, N_F fulfils certain extra conditions. We formulate these conditions here and give the proof at the end of this chapter.

Proposition 8.3: Let $p : N \longrightarrow S^1$ be a 2k-dimensional singular manifold whose boundary is a differentiable fibre bundle with classifying diffeomorphism (M,f). Let N_F be the according admissible manifold. We suppose that N_F fulfils the properties of $N'_F{}'$ in Proposition 6.11 and that $I(M,f) = 0$. Then, by a sequence of additions of handles to N_F we can obtain an admissible manifold $N'_F{}'$ which fulfils the following additional properties:

1.) $K := \mathrm{Ker}\, j_* : H_{k-1}(M;\mathbb{Q}) \longrightarrow H_{k-1}(F';\mathbb{Q})$ ia an invariant subkernel of $I(M,f) \otimes \mathbb{Q}$, where $j : M \longrightarrow F'$ is the inclusion.

2.) If $j_*x = 0$ and j_*f_*x is torsion then $j_*f_*x = 0$.

3.) If for $y \in \pi_k(N'_{F'},F'_i)$ the element $H(d^i(y)) \in H_{k-1}(F'_i)$ is torsion and if $y \circ y = 0$ then $d^i(y) = 0$.

Here $y \circ y \in \mathbb{Q}$ is defined as follows. Since $H(d^i(y))$ is torsion there exists a $n \in \mathbb{N}$ and $v \in H_k(N'_{F'})$ such that $j_*(v) = n \cdot y$, where $j: N'_{F'} \longrightarrow (N'_{F'},F'_i)$ is the inclusion. Let $y \circ y := \frac{1}{n} v \circ y \in \mathbb{Q}$. It is easy to see that this is well defined.

Before we prove this Proposition we first apply it to derive the conditions of Proposition 8.1.

<u>Proposition 8.4:</u> Let N_F be an admissible 2k-dimensional manifold which fulfils the properties of $N'_{F'}$ in Proposition 8.3 and for k even $\tau(N) = \varphi(M,f)$. Then by a finite sequence of trivial additions of handles we can replace N_F by $N'_{F'}$, such that there exists a subspace $W \subset \pi_k(N'_{F'})$ with the following properties:

$x, y \in W \Rightarrow x \circ y = 0$

$j_{i_*}|_W$ is injective, where $j_i: N'_{F'} \rightarrow (N'_{F'},F'_i)$ is the inclusion.

The rank of W is equal to rank $\pi_k(N'_{F'},F'_i) / 2$ and

$\pi_k(N'_{F'},F'_i) / j_{i_*}(W)$ is torsion free.

Remark 8.5: These properties imply that we can find a basis of W represented by embedded spheres $S^k \longrightarrow N_F$ which fulfil the properties of Proposition 8.1. Thus the proof of Propositions 8.4 and 8.3 completes the proof of Theorem 5.5 in the even-dimensional case.

Proof of Proposition 8.4: We consider the following diagram of exact sequences of homology groups with rational coefficients:

$$\begin{array}{ccccccccccc}
 & & & & 0 & & & & & & \\
 & & & & \downarrow & & & & & & \\
 & & & & H_{k+1}(N_F,\partial N_F) & & & & & & \\
 & & & & \downarrow & & & & & & \\
 & & & & H_k(\partial N_F) & & & & & & \\
 & & & & \downarrow & & & & & & \\
0 & \longrightarrow & H_k(F) & \underset{l_i}{\longrightarrow} & H_k(N_F) & \underset{j_i}{\longrightarrow} & H_k(N_F,F_i) & \underset{i}{\longrightarrow} & H_{k-1}(F) & \longrightarrow H_{k-1}(N_F) & \longrightarrow 0 \\
 & & & & \downarrow t_* & & & & & & \\
 & & & & H_k(N_F,\partial N_F) & & & & & &
\end{array}$$

The zeros in the row follow from the condition that $H_{k+1}(N_F,F_i) \cong$ $= H^{k-1}(N_F,F_{i+1}) = \{0\}$ and $H_{k-1}(N_F,F_i) = \{0\}$. The zero in the column follows from the observation that by Poincaré duality the map $H_{k+1}(N_F) \longrightarrow H_{k+1}(N_F, \partial N_F)$ corresponds to the map $H^{k-1}(N_F, \partial N_F) \longrightarrow H^{k-1}(N_F)$ which factorizes through $H^{k-1}(N_F,F_i) = \{0\}$.

We are looking for a subspace of half rank in $\pi_k(N_F,F_i)$ which goes to zero under d^i. We first do this for $H_k(N_F,F_i)$ and ∂^i. For this we compute the dimension of the rational homology $H_k(N_F,F_i)$ in terms of dimensions of $H_{k-1}(F)$ and of $H_{k-1}(N_F)$ and the rank of t.

By playing with the exact sequences in the diagram above we get:

$$\dim H_k(N_F,F_i) = \dim H_{k-1}(F) - \dim H_k(F) - 2\dim H_{k-1}(N_F)$$
$$+ \dim H_k(\partial N_F) + \operatorname{rank} t_* .$$

To compute $\dim H_k(\partial N_F) = \dim H_{k-1}(\partial N_F)$ we consider the Mayer-Vietoris sequence :

$$H_{k-1}(M) \xrightarrow{(j_*,(j\circ f)_*)} H_{k-1}(F) \oplus H_{k-1}(F) \longrightarrow H_{k-1}(\partial N_F) \longrightarrow H_{k-2}(M) \longrightarrow$$

$$\longrightarrow H_{k-2}(F) \oplus H_{k-2}(F).$$

Condition 2.) of Proposition 8.3 implies that $\operatorname{Ker} H_{k-1}(j) = \operatorname{Ker} H_{k-1}(j\circ f)$ and the condition of Proposition 6.11 implies that $H_{k-2}(j) = H_{k-2}(j\circ f)$.

Thus:

$$\dim H_{k-1}(\partial N_F) = 2\cdot \dim H_{k-1}(F) - \tfrac{1}{2}\dim H_{k-1}(M) + \dim \operatorname{Ker} H_{k-2}(j).$$

But the long exact sequence of the pair (F,M) implies:

$$\dim \operatorname{Ker} H_{k-2}(j) = \dim H_{k-1}(F,M) - \dim H_{k-1}(F) + \tfrac{1}{2}\dim H_{k-1}(M).$$

Hence:

$$\dim H_k(\partial N_F) = \dim H_{k-1}(\partial N_F) = \dim H_{k-1}(F) + \dim H_{k-1}(F,M)$$
$$= \dim H_{k-1}(F) + \dim H_k(F).$$

For $H_k(N_F,F_i)$ this implies:

$$\dim H_k(N_F,F_i) = 2(\dim H_{k-1}(F) - \dim H_{k-1}(N_F)) + \operatorname{rank} t_*.$$

Now we decompose $H_k(N_F)$ into $\operatorname{im} i_* \oplus U$ such that $t_*|_U$ is injective. Proposition 5.4 implies that $\tau(N_F) = 0$. We want to have a self annihilating subspace U_1 in U of half the dimension. By ([2], §2) this exists if and only if $\tau(N_F) = 0$ and $L(\partial N_F) = 0$ in $W(\mathbb{Q}/_{\mathbb{Z}})$, where L is the linking form on Tor $H_{k-1}(\partial N_F)$. We will show in Lemma 8.5 below that $L(\partial N_F) = 0$.

In im i_* we have two subspaces im 1_{0*} and im 1_{1*}. We choose a basis of im i_* of the following form:

$e_1,\dots,e_a$, $f_1,\dots,f_b$, $g_1,\dots,g_b$, $h_1,\dots,h_c$, where $e_1,\dots,e_a$ is a basis of im $1_{0*} \cap$ im 1_{1*}, $e_1,\dots,e_a$, $f_1,\dots,f_b$ is a basis of im 1_{0*} and $e_1,\dots,e_a$, $g_1,\dots,g_b$ is a basis of im 1_{1*}.

Let $V \subset H_k(N_F)$ be the subspace spanned by U_1 and $f_1+g_1,\dots,f_b+g_b$, $h_1,\dots,h_c$. The dimension of V is equal to $\frac{1}{2}$ rank $t_* + b + c$. A simple computation shows that $b + c = \dim H_k(\partial N_F) - \dim H_{k-1}(N_F) - \dim H_k(F)$

$$= \dim H_{k-1}(F) - \dim H_{k-1}(N_F).$$

Thus $\dim V = \frac{1}{2} \dim H_k(N_F,F_i)$. Furthermore V has the property that j_{0*} and j_{1*} restricted to V are injective and for $x,y \in V$ the intersection number $x \circ y$ vanishes.

Now we choose subspaces $V_i \subset \pi_k(N_F,F_i)$ with the following properties:

In $\pi_k(N_F,F_i) \otimes \mathbb{Q} = H_k(N_F,F_i;\mathbb{Q})$ the subspace $V_i \otimes \mathbb{Q}$ is equal to $j_{i*}(V)$ and $\pi_k(N_F,F_i)/_{V_i}$ is torsion free.

By construction $H(d^i(V_i))$ is torsion and this implies by condition 2.) of Proposition 8.3 that $d^i(V_i) = 0$. Thus there exist subspaces $V_i' \subset \pi_k(N_F)$ such that $j_{i*}|_{V_i'} : V_i' \longrightarrow V_i$ is an isomorphism and for all $x,y \in V_0', V_1'$ the intersection number $x \circ y$ vanishes.

We would be finished if $V_0' = V_1'$. To obtain this we stabilize N_F in the following sense. Let $\alpha_1,\dots,\alpha_n$ be a basis of V_0' and $\beta_1,\dots,\beta_n$ be a basis of V_1'. Now we make connected sum along the boundary of N_F within F_0 with b copies of $D^k \times S^k$ and the same with respect to F_1. Obviously this corresponds to trivial additions of handles. The resulting manifold we denote by $N'_F{}'$.

We denote the homotopy classes represented by the $\{0\} \times S^k$ by $\gamma_1,\dots,\gamma_n$ and $\delta_1,\dots,\delta_n$.

$\pi_k(N'_F{}', F_i') = \pi_k(N_F,F_i) \oplus \mathbb{Z}^n \oplus \mathbb{Z}^n$ and $\gamma_1,\dots,\gamma_n$ span a direct summand in $\pi_k(N_F,F_1)$ and vanish in $\pi_k(N_F,F_0)$ while $\delta_1,\dots,\delta_n$ span a direct summand in $\pi_k(N_F,F_0)$ and vanish in $\pi_k(N_F,F_1)$.

Let $W \subset \pi_k(N'_F{}')$ be the subspace generated by $\alpha_1 + \gamma_1,\dots,\alpha_n + \gamma_n$, $\beta_1 + \delta_1,\dots,\beta_n + \delta_n$. Then W represents a $2\cdot n$-dimensional direct summand in $\pi_k(N'_F{}',F_i')$ for i = 0 and 1, and for $x,y \in W$ the intersection number $x \circ y$ vanishes.

q.e.d.

Lemma 8.5: Let N_F^{4k} be as in Proposition 8.4. Then the linking form $L(\partial N_F)$ on Tor $H_{2k-1}(\partial N_F)$ vanis hes in $W(\mathbb{Q}/_{\mathbb{Z}})$.

Proof : We consider the following exact sequence:

$$0 \longrightarrow H_{2k-1}(F_0) \xrightarrow{i_*} H_{2k-1}(\partial N_F) \longrightarrow H_{2k-1}(F_1,M) \longrightarrow 0$$

The zero on the right side follows from the diagram

$$\begin{array}{ccc} H_{2k-2}(F_0) & \longrightarrow & H_{2k-2}(\partial N_F) \\ & \cong\searrow \quad \swarrow & \\ & H_{2k-2}(BSO) & \end{array}$$

The injectivity of Tor $H_{2k-1}(F_0) \longrightarrow$ Tor $H_{2k-1}(\partial N_F)$ follows from the assumption that $j_*x = 0$ an $j_*\ f_*\ x$ torsion implies $j_*f_*x = 0$ and from the Mayer-Vietoris sequence

$$H_{2k-1}(M) \longrightarrow H_{2k-1}(F_0) \oplus H_{2k-1}(F_1) \longrightarrow H_{2k-1}(\partial N_F)\ .$$

The next step is to show that the free part of $H_{2k-1}(F_0)$ is a direct summand of $H_{2k-1}(\partial N_F)$. By the universal coefficient theorem this is equivalent to : $H^{2k-1}(\partial N_F)/_{Tor} \longrightarrow H^{2k-1}(F_0)/_{Tor}$ is surjective or by Poincaré duality to : $H_{2k}(\partial N_F)/_{Tor} \longrightarrow H_{2k}(F_0,M)/_{Tor}$ is surjective. But this follows from the fact that $H_{2k}(\partial N_F) \longrightarrow H_{2k}(F_0,M)$ is surjective which we have shown above for F_1 instead of F_0. But the situation is completely symmetric in F_0 and F_1.

As $H_{2k-1}(F_0)$ is a direct summand in $H_{2k-1}(\partial N_F)$ we have a short exact sequence of torsion groups:

$$0 \longrightarrow \text{Tor } H_{2k-1}(F_0) \longrightarrow \text{Tor } H_{2k-1}(\partial N_F) \longrightarrow \text{Tor } H_{2k-1}(F_1,M) \longrightarrow 0$$

As Tor $H_{2k-1}(F)$ = Tor $H_{2k-1}(F,M)$ the order of Tor $H_{2k-1}(\partial N_F)$ is

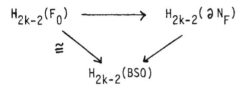

$|\mathrm{Tor}\ H_{2k-1}(F)|^2$. To show that $L(\partial N_F)$ vanishes in $W(\mathbb{Q}/\mathbb{Z})$ we have to construct a subspace $K \subset \mathrm{Tor}\ H_{2k-1}(\partial N_F)$ with $|K|^2 = |\mathrm{Tor}\ H_{2k-1}(\partial N_F)|$ (or equivalently $|K| = |\mathrm{Tor}\ H_{2k-1}(F)|$) and $K \subset K^{\perp}$.

For $x \in H_{2k-1}(F)$ we denote by x_0 and x_1 the corresponding elements in $H_{2k-1}(F_0)$ and $H_{2k-1}(F_1)$. Let K be the subspace of $H_{2k-1}(\partial N_F)$ generated by the elements $x_0 - x_1$, $x \in \mathrm{Tor}\ H_{2k-1}(F)$ and by all torsion elements in the image of $H_{2k-1}(M) \longrightarrow H_{2k-1}(\partial N_F)$.

It is not diffcult to check that $K \subset K^{\perp}$. To compute the order of K we first note that by the Mayer-Vietoris sequence above the subgroup of elements $x_0 - x_1$, $x \in \mathrm{Tor}\ H_{2k-1}(F)$ is isomorphic to $|\mathrm{Tor}\ H_{2k-1}(F)/\mathrm{im}\ H_{2k-1}(M)|$. As $H_{2k-1}(F) \longrightarrow H_{2k-1}(\partial N_F)$ is injective the images of $H_{2k-1}(M)$ in $\mathrm{Tor}\ H_{2k-1}(\partial N_F)$ and in $\mathrm{Tor}\ H_{2k-1}(F)$ are equal. Thus we are finished if the intersection of the subgroups consisting of the elements $x_0 - x_1$, $x \in \mathrm{Tor}\ H_{2k-1}(F)$ and of the image of $H_{2k-1}(M)$ in $\mathrm{Tor}\ H_{2k-1}(\partial N_F)$ is zero.

If $x \in \mathrm{Tor}\ H_{2k-1}(F)$ is non-trivial in $\mathrm{Tor}\ H_{2k-1}(F)/\mathrm{im}\ H_{2k-1}(M)$ it represents a non-trivial element in $\mathrm{Tor}\ H_{2k-1}(F,M)$. Thus there exists a $y \in \mathrm{Tor}\ H_{2k-1}(F)$ with $L(x,y) \neq 0$ as the linking form is non-singular. Obviously $L(x,y) = L(x_0 - x_1, y_0)$. On the other hand for all torsion elements z in the image of $H_{2k-1}(M) \longrightarrow H_{2k-1}(\partial N_F)$ we have $L(z,y) = 0$. thus the intersection must be zero.

q.e.d.

To complete the proof of Theorem 5.5 we now have to prove Proposition 8.3.

<u>Proof of Proposition 8.3</u> : The main tool for the first step is the following

Lemma 8.6 : Given $x \in H_{k-1}(M)$ with $x \circ x = 0$ and $y \in \pi_{k-1}(F)$ with $\nu_*(y) = 0$ in $\pi_{k-1}(BSO)$ such that $j_*(x) = H(y)$. Then we can represent 2y by an embedding $S^{k-1} \times D^k \longrightarrow F$ such that if F' is obtained by surgery with this embedding the following holds:

1.) $j'_*(2x) = 0$

2.) If for $z \in H_{k-1}(M)$ $x \circ z = 0$ and $j_*(z) = 0$ then $j'_*(z) = 0$.

Proof : The second statement is clear. For if $j_*(z) = 0$ then there exists a simplicial chain c in F with $dc = z$. c is a relative cycle in (F,M) and $0 = z \circ x = c \circ j_* x = c \circ H(y)$. By the Whitney trick we can choose c so that $c \cap S^{k-1} = \emptyset$.

For the first statement we begin by representing y by an embedding $S^{k-1} \times D^k \longrightarrow F$ denoted by U. We have the following diagram of exact sequences:

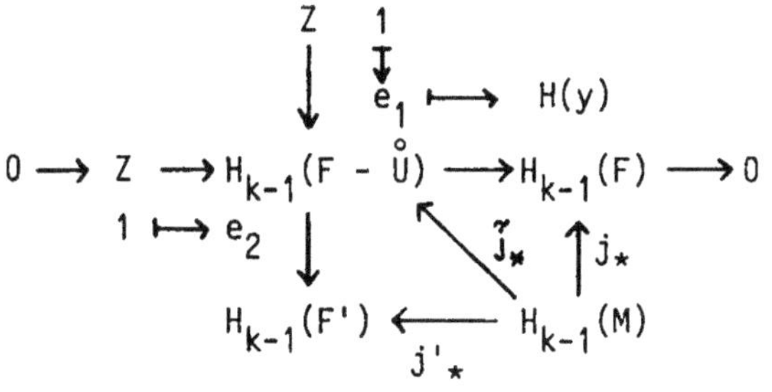

As $H(y) - j_*(x) = 0$ there exists a $s \in Z$ such that $e_1 + se_2 - \tilde{j}_*(x) = 0$ in $H_{k-1}(F - \mathring{U})$, where e_1 and e_2 are represented by $S^{k-1} \times \{*\}$ and $\{*\} \times S^{k-1}$. If k is odd this implies $0 = (e_1 + se_2 - x) \circ (e_1 + se_2 - x) = 2s$. Thus $s = 0$ and $j'_*(x) = 0$.

If k is even we know from ([10] , p. 105) that we can change s by an arbitrary even number if we twist the embedding appropriately. So, we are finished if s is even. We will show that we can achieve s even by passing from y to 2y. Let us denote the given embedding $S^{k-1} \times D^k \longrightarrow F$ by 1.

Let $g: S^{k-1} \longrightarrow S^{k-1}$ be the map $(z,v) \longmapsto (z^2,v)$ where we consider S^{k-1} as a subspace of $\mathbb{C} \times \mathbb{R}^{k-1}$. Consider the embedding

$1': S^{k-1} \times D^k \longrightarrow \mathring{F}$ given by $(x,y) \longmapsto 1(g(x), \frac{1}{2}x + \frac{1}{4}y)$.

$1(S^{k-1} \times D^k)$

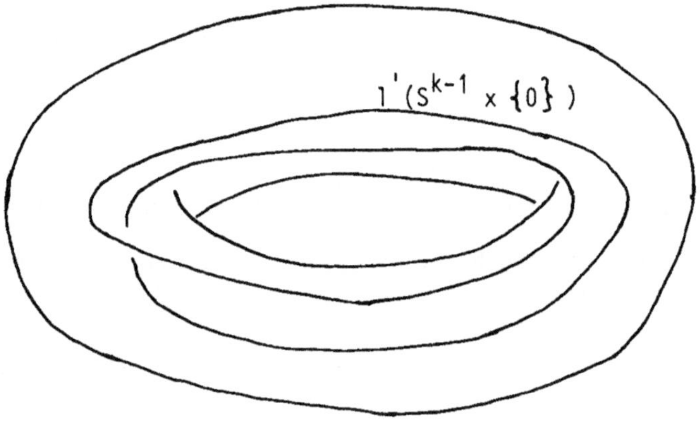

In $F - 1'(S^{k-1} \times \mathring{D}^k)$ we have:

$e_1' = 2\,e_1 + e_2$

$2\,e_2' = e_2$

Thus $1'$ represents $2 \cdot y$ in $H_{k-1}(F)$ and in $H_{k-1}(F-\mathring{U}')$ we have:

$e_1' + 2(2 \cdot s-1)\,e_2' - \tilde{j}_*'(y) = 2(e_1 + se_2) - \tilde{j}_*(y) = 0.$

q.e.d.

With this Lemma we proceed in the proof of Proposotion 8.3 as follows.

Since $I(M,f) = 0$ there exists an invariant subkernel $K \subset H_{k-1}(M)$. We distinguish between two cases.

1.) $f_*(K \cap \text{Ker } j_*) \otimes \mathbb{Q} \neq (K \cap \text{Ker } j_*) \otimes \mathbb{Q}$

Then there exists $x \in K \cap \text{Ker } j_*$ with $j_* f_*(x) \neq 0$ in $H_{k-1}(F;\mathbb{Q})$.

The following diagram

$$\begin{array}{ccc} & & H_k(N_F,F_0) = \pi_k(N_F,F_0) \\ & & \downarrow \\ H_{k-1}(M) & \longrightarrow & H_{k-1}(F) \\ \downarrow & & \downarrow \\ H_{k-1}(BSO) & \overset{\cong}{\longleftarrow} & H_{k-1}(N_F) \end{array}$$

implies that $j_*(x-f_*(x))$ maps to zero in $H_{k-1}(N_F)$ and thus there exists a $y \in \pi_{k-1}(F)$ with $H(y) = j_*(x-f_*(x))$ and $\nu_*(y) = 0$. As $j_*(x) = 0$ the element $f_*(x)$ fulfils the conditions of Lemma 8.6. By addition of handles corresponding to surgery on Y as in the Lemma we enlarge the dimension of $(K \cap \text{Ker } j_*) \otimes \mathbb{Q}$.

Thus after a finite number of such additions of handles we can assume

2.) $f_*(K \cap \text{Ker } j_*) \otimes \mathbb{Q} = (K \cap \text{Ker } j_*) \otimes \mathbb{Q}$.

As the intersection form on $H_{k-1}(M;\mathbb{Q})$ is non-singular, for each non-trivial element $x \in H_{k-1}(M;\mathbb{Q})$ which is not contained in the subgroup generated by K and $\text{Ker } j_*$ there exists a $y \in K \cap \text{Ker } j_*$ with $x \circ y \neq 0$. This implies that for all $x \in \text{Ker } j_*$ the element f_*x is contained in the subgroup generated by K and $\text{Ker } j_*$. For otherwise there would be a $y \in K \cap \text{Ker } j_*$ with $f_*x \circ y \neq 0$ which leads to a contradiction as we know from $f_*(K \cap \text{Ker } j_*)$ $= K \cap \text{Ker } j_*$ that $y = f_*z$ with $z \in K \cap \text{Ker } j_*$ and thus $f_*x \circ y = f_*x \circ f_*z$ $= x \circ z = 0$.

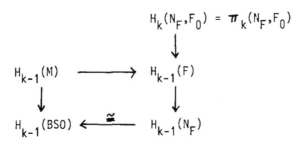

Now, either $f_*(\mathrm{Ker}\ j_*) = \mathrm{Ker}\ j_*$ and we would be finished or there exists a $x \in \mathrm{Ker}\ j_*$ with $j_*f_*x \neq 0$. As $f_*(x) \in \langle K, \mathrm{Ker}\ j_* \rangle$ we know that $j_*f_*x = j_*y$ for some $y \in K$. As above we can represent $2j_*y$ in F by an embedding $S^{k-1} \times D^k$ so that after surgery with this embedding $j'_*y = 0$ and as before $j'_*(K \cap \mathrm{Ker}\ j_*) = 0$. Thus again we can enlarge the dimension of $(K \cap \mathrm{Ker}\ j_*) \otimes \mathbb{Q}$ which finally must lead to the situation that $\mathrm{Ker}\ j_* \otimes \mathbb{Q}$ is an invariant subkernel.

For the proof of the second statement we consider a $x \in H_{k-1}(M)$ with $j_*x = 0$ and $j_*f_*x \neq 0$ but torsion. As above we know that then j_*f_*x can be represented by an embedding $S^{k-1} \times D^k \longrightarrow F$. With the same notations and exact sequences as in the proof of Lemma 8.6 we see that

$$r\, e_1 + s\, e_2 = 0$$

in $H_{k-1}(F - \mathring{U})$, where r is the order of j_*f_*x in $H_{k-1}(F)$. If k is odd the same argument as in the proof of Lemma 8.6 shows that s is 0 and thus we can kill j_*f_*x by surgery.

If k is even the number r/s in $\mathbb{Q}/2\mathbb{Z}$ $(k \neq 2,4,8)$ is the strong self linking number as defined before Proposition 6.11 . For $k = 2,4,8$ the number $r/s \in \mathbb{Q}/\mathbb{Z}$ is the ordinary self linking number. Again as in the proof of Lemma 8.6 we are finished if r/s is zero in $\mathbb{Q}/2\mathbb{Z}$ for $k \neq 2,4,8$ or in $\mathbb{Q}/\mathbb{Z}$ for $k = 2,4,8$. For one knows (compare [10], p.105) that one can change s by an arbitrary multiple of 2r (r, if $k = 2,4,8$) if we twist the embedding appropriately.

By the last condition of Proposition 6.11 we have for $k \neq 2,4,8$ a $y \in \pi_{k-1}(F)$ with $\nu_*(y) = 0$, $H(y) = 0$ and $g(y) = 1$. Thus it is enough to show that r/s is 0 or 1 in $\mathbb{Q}/2\mathbb{Z}$ or equivalently that the ordinary self linking number of j_*f_*x vanishes. For then in the last case we can also achieve that it is 0 in $\mathbb{Q}/2\mathbb{Z}$ by replacing the given embedding by the connected sum with an embedding representing y.

The ordinary self linking number doesn't change if we embed F into $F \cup_f -F$. But $j_* x = 0$ in $H_{k-1}(F)$ implies that $j_* f_* x$ is trivial in $H_{k-1}(F \cup_f -F)$.

To prove the last statement we need some properties of the winding number of embeddings.

Let F be a 2k-1-dimensional manifold with boundary and l_1, l_2 disjoint embeddings of closed oriented manifolds of dimension k-1 in $\mathring{F}$ which we denote by A and B. We suppose that the homology classes $[A]$ and $[B]$ represented by these submanifolds are torsion. This implies that there exists a $n \in \mathbb{N}$ and a class $v \in H_k(F,A;\mathbb{Z})$ with $\partial v = n \cdot [A]$. The intersection number between v and $[B]$ is defined and we use it to define the <u>winding number</u> of l_1 and l_2 by

$$L(l_1,l_2) := \frac{1}{n}(v \circ [B]) \in \mathbb{Q}.$$

It is obvious that this number is well defined. It depends on the embedding and not only on the homology classes represented by A and B.
Now let W be a 2k-dimensional manifold with boundary which contains F as a submanifold of the boundary. In general for elements in $H_k(W,F;\mathbb{Z})$ there is no intersection number defined. But under additional assumptions this is possible. Let $z_1, z_2 \in H_k(W,F;\mathbb{Z})$ be such that ∂z_i is torsion, where $\partial : H_k(W,F;\mathbb{Z}) \longrightarrow H_{k-1}(F;\mathbb{Z})$ is the boundary operator. Then we can define the <u>intersection number</u> $z_1 \circ z_2$ as follows. As ∂z_1 is torsion there exists a $m \in \mathbb{N}$ and $w \in H_k(W;\mathbb{Z})$ with $i_* w = m \cdot z_1$, $i: W \longrightarrow (W,F)$ the inclusion.
$z_1 \circ z_2 := \frac{1}{m}(w \circ z_2) \in \mathbb{Q}$.

Again it's easy to see that this intersection number is well defined.

We need the following well known Lemma giving a connection between the winding number and the intersection number. For completeness we give a short proof.

Lemma 8.7: Let (W,F) be as above. Let $s_1 : (C_1^k, \partial C_1) \longrightarrow (W, \mathring{F})$ and $s_2 : (C_2^k, \partial C_2) \longrightarrow (W, \mathring{F})$ be disjoint embeddings meeting the boundary of W transversally. We suppose that the homology classes $[s_1(\partial C_1)]$ and $[s_2(\partial C_2)]$ are torsion then:

$$[s_1(C_1), s_1(\partial C_1)] \circ [s_2(C_2), s_2(\partial C_2)] = - L(s_1|\partial C_1, s_2|\partial C_2)$$

Proof: Choose a triangulation of (W,F) such that $s_1(C_1)$ and $s_2(C_2)$ are subcomplexes. We will use simplicial homology. Let $n \cdot [s_1(\partial C_1)] = 0$. Then there exists a cycle $v \in Z_k(F, s_1(\partial C_1))$ with $\partial v = n [s_1(\partial C_1)]$ considered as a cycle. $w := n \cdot [s_1(\partial C_1)] - v$ is a cycle in $Z_k(W)$ which represents the homology class $n \cdot [s_1(C_1), s_1(\partial C_1)]$ in $H_k(W,F;\mathbb{Z})$. By definition we have:

$$[s_1(C_1), s_1(\partial C_1)] \circ [s_2(C_1), s_2(\partial C_2)] = \frac{1}{n} (w \circ s_2 [C_2]) =$$

$$= - \frac{1}{n} (v \circ [s_2(\partial C_2)]), \text{as } s_1(C_1) \cap s_2(C_2) = \emptyset.$$

But the last expression is equal to $- L(s_1|\partial C_1, s_2|\partial C_2)$.

q.e.d.

Now we can prove the last statement. Let $z \in \pi_k(N_F, F_i)$ be so that $H d^i z$ is torsion and $z \circ z = 0$. We represent z by an embedding $(D^k, S^{k-1}) \longrightarrow (N, \mathring{F}$ meeting F_i transversally. Let $l : S^{k-1} x D^k \longrightarrow F_i$ be the restriction of a tubular neighborhood of this embedding. Let $N'_F{}'$ be obtained from N_F by addition of handles with l.

As $H\,d^i(z)$ is torsion $\mathrm{Ker}\, j_* \otimes \mathbb{Q} = \mathrm{Ker}\, j'_* \otimes \mathbb{Q}$. This follows by a similar argument as the proof of the second statement of Lemma 8.6.

We will show now that $l(\{*\} \times S^{k-1}) \subset F'$ has infinite order in $H_{k-1}(F')$. For this we denote $l(S^{k-1} \times D^k)$ by U and consider the well known diagram

$$\begin{array}{ccccccccc}
 & & & & \mathbb{Z} \quad 1 & & & & \\
 & & & & \downarrow \quad \downarrow e_1 & & & & \\
0 & \longrightarrow & \mathbb{Z} & \longrightarrow & H_{k-1}(F-\mathring{U}) & \longrightarrow & H_{k-1}(F) & \longrightarrow & 0 \\
 & & 1 & \longmapsto & e_2 & & & & \\
 & & & & \downarrow & & & & \\
 & & & & H_{k-1}(F') & & & & \\
 & & & & \downarrow & & & & \\
 & & & & 0 & & & &
\end{array}$$

If $H\,d^i(z)$ has order n then we have the equation $ne_1 + \lambda e_2 = 0$ in $H_{k-1}(F-\mathring{U})$, where λ is the order of $l(\{*\} \times S^{k-1})$ in $H_{k-1}(F')$. On the other hand λ/n is the linking number $L(l(S^{k-1} \times \{*\}, l(S^{k-1} \times \{0\}))$, where $\{*\} \in S^{k-1} = \partial D^k$. But Lemma 8.7 implies that this number is equal to $- z \circ z = 0$. Thus $\lambda = 0$ or $l(\{*\} \times S^{k-1})$ has infinite order in $H_{k-1}(F')$.

Now there are two possibilities. If $d^i(z)$ has infinite order then dim $\mathrm{Ker}\, H' \otimes \mathbb{Q} < \dim \mathrm{Ker}\, H \otimes \mathbb{Q}$. Thus after finitely many additions of handles as above we can assume that $d^i(z)$ is torsion. But if $d^i(z)$ is a non-trivial torsion element then, as $l(\{*\} \times S^{k-1})$ has infinite order in $\pi_{k-1}(F')$, it follows that $|\mathrm{Tor}\, \pi_{k-1}(F')| < |\mathrm{Tor}\, \pi_{k-1}(F)|$. Thus we can achieve our last statement by a sequence of such additions of handles.

q.e.d.

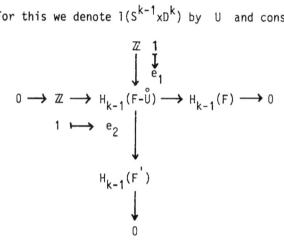

§ 9   Bordism of diffeomorphisms on manifolds with additional normal

structures like Spin-, unitary structures or framings; orientation

reversing diffeomorphisms and the unoriented case.

Additional normal   structures on an oriented manifold are specefied by

giving a factorization of the classifying map of the stable normal bundle

$\nu : M \longrightarrow BSO$ over a fibration B over BSO:

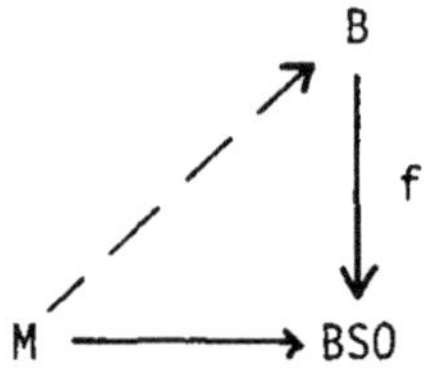

For instance one has fibrations of B Spin or BU over BSO and a factoriza-

tion of the classifying map over B Spin or BU corresponds to a reduction

of the structure group to Spin or U.

More precisely, let $(B,f) = (B_k, f_k, g_k)$ be sequence of fibrations

$f_k : B_k \longrightarrow BSO(k)$ and maps $g_k : B_k \longrightarrow B_{k+1}$ such that the diagrams

$$
\begin{array}{ccc}
B_k & \xrightarrow{\;g_k\;} & B_{k+1} \\
\big\downarrow{\scriptstyle f_k} & & \big\downarrow{\scriptstyle f_{k+1}} \\
BSO(k) & \longrightarrow & BSO(k+1)
\end{array}
$$

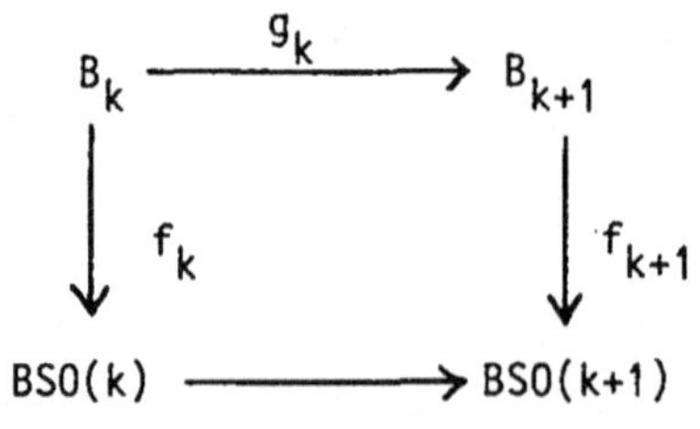

commute.

Then Lashof has defined a (B,f)-structure on an oriented manifold M as

follows ([27] , see also [43] , p. 14 ff.) Let $\nu_k : M \longrightarrow BSO(k)$ be the

classifying map of the stable oriented normal bundle of M, obtained by embedding M into R^{n+k} and then taking the Gauß map. Then a (B,f)-structure on M is a sequence of homotopy liftings $\eta_k : M \longrightarrow B_k$ of ν_k commuting with g_k.

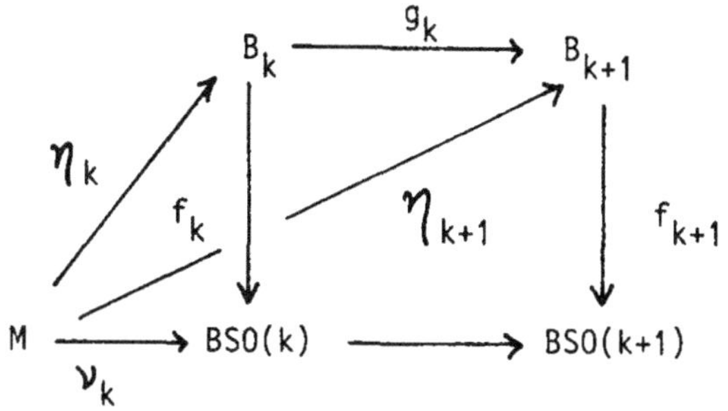

We now give some additional examples of (B,f)-structures (compare Stong [43], § IV). We already have mentioned the structure groups to some special classical groups. This corresponds to the (B,f)-structures given by the fibrations B Spin(k), BU(k), B SU(k) or B(1) over BSO(k) where 1 denotes the trivial group. In the last case a (B,f)-structure corresponds to a framing on M. Other examples are given by the r-connected cover over BSO(k). In this case one gets examples of manifolds which have such a structure by r-connected manifolds and for $r < \frac{1}{2}$ dim M, the corresponding bordism groups are actually the same.

There is an obvious way to introduce the bordism group of manifolds with (B,f)-structure ([43], p. 17). In this chapter we want to consider diffeomorphisms on manifolds with (B,f)-structure which preserve this structure and compute the corresponding bordism groups. Before we define the concept of a diffeomorphism preserving a (B,f)-structure we want to study the situation which should occur for such a diffeomorphism. Such a diffeomorphism f on M should have the property that the mapping torus M_f admits a (B,f)-structure which extends the given (B,f)-structure on M.

In the case of diffeomorphism on oriented manifolds the fundamental step in the computation of the corresponding bordism group was to consider the problem under which conditions a manifold whose boundary is a differentiable fibre bundle over S^1 is bordant to a manifold such that the fibre bundle can be extended. The same question can be stated for manifolds with (B,f)-structure and it has the same answer if B fulfils certain properties.

Theorem 9.1: Let (N,η) be an oriented (B,f)-manifold of dimension $m > 4$ and $p: N \longrightarrow S^1$ a map such that $p|_{\partial N}$ is a differentiable fibre bundle. Let $f: M \longrightarrow M$ be the classifying diffeomorphism of $p|_{\partial N}$. We assume that B is 1 - connected. Then (N,η,p) is bordant rel. boundary as a singular (B,f) - manifold to a differentiable fibre bundle over S^1 whose restriction to the boundary is the given bundle if and only if:

a) m is odd.

b) $m \equiv 2(4)$ and $I(M,f) = 0$.

c) $m \equiv 0(4)$ and $I(M,f) = 0$ and $\tau(N) = \varphi(M,f)$.

Proof of Theorem 9.1: We will see that the proof of Theorem 5.5 extends to this situation. The first step for this is to check under which assumptions the modifications used there can be applied so that the (B,f)-structures are respected. The modifications are surgery and addition and subtraction of handles. Obviously there is no problem with subtraction of handles. For surgery and addition of handles the necessary and sufficient condition to preserve a given (B,f)-structure is that the embedded spheres on which one is performing surgery or addition of handles must be mapped to zero in the corresponding homotopy group of B. This follows from ([23], Lemma 6.2) where it is proved for framed manifolds.

Now we discuss the different statements which were used in the proof of Theorem 5.5 and ask whether they extend to the case of (B,f)-manifolds. The first is Lemma 6.4 which extends as B is assumed to be 1-connected. The next is Lemma 6.6 which extends as elements in the image of d^i are mapped to zero in B. This implies that Proposition 6.10 holds in the (B,f)-situation and as we only use subtraction of handles in § 7 this shows that the odd-dimensional result is true.

Before we continue with the discussion of the even - dimensional case the following observation: Our aim is to modify N_F, which is already assumed to be 1 - connected, by surgeries and addition and subtraction of handles within B into a h - cobordism. The map $\tilde{\nu}: N_F \longrightarrow B$ factors through $id: N_F \longrightarrow N_F$, which is a N_F - structure on N_F. Obviously it is enough to make the modification of N_F within N_F instead of B. We make this remark as it shows that it is enough to prove the theorem if B is a finite CW - complex which we will assume in the following.

Now we continue with the even - dimensional case. In the formulation of Proposition 6.11 we have to replace $\nu: F \longrightarrow BSO$ by $\tilde{\nu}: F \longrightarrow B$. Furhtermore the additional statement about the strong self linking number can only be true if there exists any closed (2k-1)- dimensional B - manifold X which is zero bordant and has a $x \in \pi_{k-1}(X)$ with $H(x) = 0$, $\tilde{\nu}(x) = 0$ and $\varphi(x) = 1$. Thus we change the formulation of Proposition 6.11 by assuming that B is finite and we require the existence of an element with non-trivial strong self linking number only if such a B - manifold X exists. Then the proof of Proposition 6.11 extends without any problem.

There is no problem with Proposition 8.1 which only uses subtraction of handles. In the proof of Proposition 8.3 one has to reformulate Lemma 8.6 for B instead of BSO as above and with this the proof of statement 1.)

extends if we replace BSO by B at all places. In the proof of statement 2.) we have made use of the existence of an element $x \in \pi_{k-1}(F)$ with $\rho(x) = 1$, $\tilde{\gamma}_*(x) = 0$ and $H(x) = 0$. By Proposition 6.11 we can assume this now only if there exists some closed B - manifold X with the properties above.

If we look into the proof of statement 2.) of Proposition 8.3 we see that for $x \in H_{k-1}(M)$ with $j_*x = 0$ we can represent j_*f_*x by a sphere $S^{k-1} \subset F$. Now there are two possibilities: either $\rho(S^{k-1}) = 0$ and we can kill j_*f_*x by surgery or $\rho(S^{k-1}) = 1$ $(k \neq 2,4,8)$. In the last case we consider $X = F \cup_f -F$. As $j_*x = 0$ the sphere $S^{k-1} \subset F$ represents zero in $H_{k-1}(X)$. On the other hand $\rho(S^{k-1}) = 1$. So, in this case we have a closed zero bordant B - manifold X with the properties needed in the modified formulation of Proposition 6.11. Thus in any case the proof of Proposition 8.3 can be extended.

There is no problem with the extension of Proposition 8.4 .

q.e.d.

<u>Corollary 9.2</u> : Let B be 1 - connected. Then a closed m - dimensional (B,f) - manifold ($m \geq 5$) is (B,f) - bordant to a fibration over S^1 if and only if the signature vanishes.

<u>Remarks 9.3</u> :

1.)For unitary manifolds this Corollary is known ([5] ,Theorem 5). It follows from the computation of the unitary SK - groups. On the other hand we will use the Corollary in the next chapter to compute the SK - groups for more general (B,f) - manifolds.

2.) For framed mainifolds the Corollary is a simple consequence of the result that every framed manifold except an Arf-invariant 1 manifold is framed bordant to a homotopy sphere which obviously is framed bordant to a mapping torus with fibre the standard sphere.

If there exists a framed Arf-invariant 1 manifold in dim 4k+2 then one can construct a representative fibred over S^1 as follows. Let S be the sphere bundle of the tangent bundle of S^{2k+1} and D an embedded disk in S. Then $S - \mathring{D}$ can be obtained by plumbing the trivial bundle over S^{2k} with fibre D^{2k+1} with a disk bundle over S^{2k+1} that corresponds (after addition of a 1-dim trivial bundle) to the tangent bundle of S^{2k+1}. Then we can use the classifying map of the tangent bundle of S^{2k+1}, $\alpha : S^{2k} \longrightarrow SO(2k+1)$ to get a diffeomorphism f on $S - \mathring{D}$ by twisting the trivial bundle in $S - \mathring{D}$ with α and extending it to the rest by identity. $f|_{\partial(S-\mathring{D})=S^{4k}}$ corresponds to a homotopy sphere Σ^{4k+1} which can easily be identified with the Kervaire sphere [23]. Now the existence of a closed framed Arf-invariant 1 manifold is equivalent to $\Sigma \cong S^{4k+1}$ or to f isotopic to the identity. Then we can extend f to a diffeomorphism f on S which preserves the framing of S up to homotopy. Thus the mapping torus of f is a framed manifold which has Arf-invariant 1.

Now we want to introduce the bordism groups of (B,f)-structure preserving diffeomorphism on (B,f)-manifolds. It is natural to require that a diffeomorphism g on a (B,f)-manifold (M,η) preserves η up to homotopy, that means that $\eta \circ g$ is homotopic to η. But it is clear that this information about g is not enough to introduce the corresponding bordism group as the bordism relation would not be transitive. For if (W,η, F) is a bordism between (M_1, η_1, f_1) and (M_2, η_2, f_2) and (W', η', F') is a bordism between (M_2, η_2, f_2) and (M_3, η_3, f_3) then in general $F \cup F'$ doesn't preserve the (B,f)-structure on $(W, \eta) \cup (W', \eta')$. The reason is that the homotopies between η_2 and $\eta_2 \circ F|_{M_2}$ and between η_2 and $\eta_2 \circ F'|_{M_2}$ might not coincide.

Thus it is necessary to include the homotopy into the definition of a (B,f)-structure preserving diffeomorphism.

Definition 9.4: Let (M, η) be a (B,f)-manifold. A (B,f)-structure preserving diffeomorphism on (M, η) is a pair (g,h) where g is a diffeomorphism on M and h is a homotopy between η and $\eta \circ g$. More precisely h is a family of homotopies h_k from $M \times I \longrightarrow B_k$ between η_k and $\eta_k \circ g$ such that for all t the sequence $(h_k)_t$ is a (B,f)-structure on M.

We denote a (B,f)-structure preserving diffeomorphism by (M, η, g, h). There is an obvious bordism relation for these diffeomorphisms. The corresponding bordism group of m-dimensional diffeomorphisms is denoted by $\Delta_m^{(B,f)}$.

Remark 9.5: The information in (M, η, g, h) given by h is equivalent to a (B,f)-structure on the mapping torus M_h extending the given (B,f)-structure on M. It is sometimes easier to think of h in this manner.

In particular this implies that we have a well defined map $\Delta_m^{(B,f)} \to \Omega_{m+1}^{(B,f)}$ given by the mapping torus. As for oriented manifolds the vanishing of the mapping torus in $\Omega_{m+1}^{(B,f)}$ and the vanishing of $[M, \eta]$ in $\Omega_m^{(B,f)}$ gives us a (B,f)-manifold N and a continous map $p: N \longrightarrow S^1$ such that $p|_{\partial N}$ is the projection of the mapping torus onto S^1. Thus if B is as in Theorem 9.1 and $m+2 \neq 0(4)$ we see that the bordism class of (M,η) and of the mapping torus and the isometricstructure are the only invariants for the bordism class $[M,\eta,f,h]$. For $m+2 = 0(4)$ and N a singular B - manifold in S^1 with $\partial N = M_f$ and $I(M,f) = 0$ we could conclude that $(M,\eta,f,h,)$ is zero bordant if we would know that $\tau(N) = \varphi(M,f)$. For oriented manifolds we always could achieve this by adding to N an appropriately multiple of the complex projective space. In the general situation one only can conclude that the kernel of the three invariants is a subgroup of $Z/_{\tau(B,m+2)Z}$, where $\tau(B,m)$ is the smallest positive signature of a closed m - dimensional B - manifold.

Corollary 9.6 : Let $m \geq 3$ and B be 1-connected.

The homomorphism

$\Delta_m^{(B,f)} \longrightarrow \Omega_m^{(B,f)} \oplus \Omega_{m+1}^{(B,f)}$ is injective for m odd and the homomorphism

$\Delta_m^{(B,f)} \longrightarrow \Omega_m^{(B,f)} \oplus \Omega_{m+1}^{(B,f)} \oplus W_{(-1)^{m/2}}(Z;Z)$ is injective for

$m = 0(4)$ and has kernel a subgroup of $Z/_{\tau(B,m+2)Z}$ for $m = 2(4)$.

Next, we want to study the image of these homomorphisms. It is obvious that the map $\Delta_m^{(B,f)} \longrightarrow \Omega_m^{(B,f)}$ is surjective and we know from Corollary 9.2 that the image of the map $\Delta_m^{(B,f)} \longrightarrow \Omega_{m+1}^{(B,f)}$ is the kernel of the signature.

To study the image of the isometric structure I: $\Delta_{2m}^{(B,f)} \longrightarrow W_\epsilon(\mathbb{Z};\mathbb{Z})$ we first consider the case of framed manifolds as those manifolds have (B,f)-structures for all other (B,f). There is an obvious restriction on the image of I: $\Delta^{fr}_{4m} \longrightarrow W_+(\mathbb{Z};\mathbb{Z})$ as the intersection form of a 4m-dim framed manifold is always even. We denote the subgroup in $W_+(\mathbb{Z};\mathbb{Z})$ with even bilinear form by $W_+^{even}(\mathbb{Z};\mathbb{Z})$. A second restriction comes from the fact that the signature of a framed manifold vanishes. We denote the kernel of the signature in $W_+^{even}(\mathbb{Z};\mathbb{Z})$ by $\hat{W}_+^{even}(\mathbb{Z};\mathbb{Z})$.

Proposition 9.7: The homomorphisms

$$\Delta^{fr}_{4k} \longrightarrow \hat{W}_+^{even}(\mathbb{Z};\mathbb{Z}) \text{ and}$$

$$\Delta^{fr}_{4k+2} \longrightarrow W_-(\mathbb{Z};\mathbb{Z}) \text{ are surjective.}$$

Proof: As every element in $W_+^{even}(\mathbb{Z};\mathbb{Z})$ or in $W_-(\mathbb{Z};\mathbb{Z})$ is equivalent to an element whose bilinear form is the intersection form of

$$r \cdot (S^n \times S^n) = S^n \times S^n \# S^n \times S^n \# \dots \# S^n \times S^n$$

for an appropriate r [33] it is enough to show that every isometry of the intersection form of $r(S^n \times S^n)$ can be realized by a diffeomorphism preserving the standard framing.

Let h be such an isometry. Then it is well known that h can be realized by a diffeomorphism g on $M - D^{2n}$, $M = r(S^n \times S^n)$ ([48] , [47]). For $n > 2$ the idea is to choose spheres $S^n_i \subset M - \mathring{D}^{2n}$ representing a basis of the middle homology and to choose spheres $(S^n_i)'$ representing the image of this basis under h. Then f maps S^n_i to $(S^n_i)'$ and one can extend this to tubular neighborhoods of S^n_i and $(S^n_i)'$ to get a diffeomorphism f from $M - \mathring{D}^{2n}$ into $M - \mathring{D}^{2n}$. By choosing the tubular neighborhoods in such a way, that they respect the framing on M one gets an f which preserves the framing on $M - \mathring{D}^{2n}$.

The next step is to extend f to M. For this consider $f|_{S^{2n-1}}$. This corresponds to a homotopy sphere Σ and f gives us a framing preserving diffeomorphisms $M \# \Sigma = (M-\mathring{D}^{2n}) \cup_{f|_{S^{2n-1}}} D^{2n} \longrightarrow M$. But then $M \# \Sigma$ and M are framed bordant and this implies that $\Sigma \in bP_{2n+1} = \{0\}$.

Thus f can be extended to a diffeomorphism f from M into M. The obstruction to the extension of the homotopy between the framing on $M - \mathring{D}^{2n}$ and the framing twisted by $f|_{M-\mathring{D}^{2n}}$ to a homotopy on M is an element of $\widetilde{KO}(S^{2n+1}) = \pi_{2n}(SO)$. By construction this element is in the kernel of the J-homomorphism $J: \pi_{2n}(SO) \longrightarrow \pi_{2n}{}^{s}$ which is injective by Adams [1].

q.e.d.

The next step is to compare $W_+^{even}(\mathbb{Z};\mathbb{Z})$ and $W_+(\mathbb{Z};\mathbb{Z})$. It is well known that the signature of an even form is divisible by 8 (compare [33]) and there is a second condition, for an isometric structure g of an even form fulfils rank $(1-g)$ = rank $(1-g \otimes \mathbb{Z}_2)$ mod 2. In the case of a diffeomorphism this corresponds to the de Rham invariant. As before we call rank $(1-g)$ - rank $(1-g \otimes \mathbb{Z}_2)$ mod $2 \in \mathbb{Z}_2$ the de Rham invariant of the isometric structure.

Proposition 9.8: The following sequence is exact:

$$0 \longrightarrow W_+^{even}(\mathbb{Z};\mathbb{Z}) \longrightarrow W_+(\mathbb{Z};\mathbb{Z}) \longrightarrow \mathbb{Z}/_8 \oplus \mathbb{Z}/_2 \longrightarrow 0$$

$$[V,s,g] \longmapsto (\tau(V) \bmod 8,\ \text{de Rham}(g))$$

This Proposition is proved in the Appendix, Corollary 2.

Now we are prepared for a description of $\Delta_m^{(B,f)}$ for $m \geq 3$.

Theorem 9.9 : Let B be 1-connected , $m \geq 3$.

For m odd there is an isomorphism

$$\Delta_m^{(B,f)} \longrightarrow \Omega_m^{(B,f)} \oplus \hat{\Omega}_{m+1}^{(B,f)}\ ,\ \hat{\Omega}_{m+1}^{(B,f)} \text{ the kernel of the signature.}$$

For $m = 2(4)$ there is a surjective map

$$\Delta_m^{(B,f)} \to \Omega_m^{(B,f)} \oplus \Omega_{m+1}^{(B,f)} \oplus W_-(Z;Z) \text{ with kernel a subgroup of } Z/_{\tau(B,m+2)Z}.$$

For $m = 0(4)$ there is an exact sequence

$$0 \to \Delta_m^{(B,f)} \to W_+(Z;Z) \oplus \Omega_m^{(B,f)} \oplus \Omega_{m+1}^{(B,f)} \to Z \oplus Z_2 \to 0$$

Proof : From Corollary 9.2 we know that $\Delta_m^{(B,f)} \to \Omega_{m+1}^{(B,f)}$ is surjective. From this and Corollary 9.6 and Proposition 9.7 one can conclude the first two statements in the same manner as for oriented manifolds.

In the exact sequence the map into $Z + Z_2$ is given by $(x,[M],[N]) \longmapsto$ (sign x - sign M, de Rham x - $w_2 w_{m-1}(N)$). We only have to show that if $(x, [M], [N])$ goes to 0 in $\mathbb{Z} \oplus \mathbb{Z}_2$ then it comes from $\Delta_{4m}^{(B,f)}$. But as $\Delta_{4m}^{(B,f)} \longrightarrow \Omega_{4m+1}^{(B,f)}$ is surjective it is enough to show that if (x,0,0) goes to 0 in $\mathbb{Z} \oplus \mathbb{Z}_2$ then x is the isometric structure of a diffeomorphism. From Proposition 9.8 we know that x is equivalent to an element in $W_+^{even}(\mathbb{Z};\mathbb{Z})$.

From the proof of Proposition 9.7 we know that there is a diffeomorphism f on $r(S^{2m} \times S^{2m})$ whose isometric structure is x, where $r(S^{2m} \times S^{2m})$ has a null-bordant framing and f respects this framing. We would be finished if we could choose f such that the framing of the fibre of the mapping torus extends to a null-bordant framing on the mapping torus.

We are free to change f by composition with a diffeomorphism which is the identity outside a disc D^{4n} in $r(S^{2n} \times S^{2n})$. This changes the mapping torus by the connected sum with the corresponding homotopy sphere ([9] , Lemma 1) and we are free to choose the framing on the homotopy sphere. Because every framed 4n-manifold is framed bordant to a homotopy sphere ([23] , Lemma 7.3) we may choose f so that the mapping torus is framed null bordant.

q.e.d.

There are two other types of bordism of diffeomorphisms of general interest: Orientation reversing diffeomorphisms on oriented manifolds and diffeomorphisms on unoriented manifolds. The second is implicitely contained in Quinn's work [38] and we state the result later.

We denote the bordism group of orientation reversing diffeomorphisms on oriented manifolds by Δ_m^- . Note that this group consists only of 2-torsion, for an orientation reversing diffeomorphism (M,f) is diffeomorphic to (-M,f), the diffeomorphism is given by f. There is a natural definition of an admissible manifold for an orientation reversing diffeomorphism:

An oriented manifold N_F is an admissible manifold for the orientation reversing diffeomorphism (M,f) if $\partial(N_F) = F \cup_f (+F)$, where F is an oriented manifold with $\partial F = M$.

By slight modifications of the arguments in § 6 one can introduce addition and subtraction of handles for such an admissible manifold. Furthermore in analogy to Proposition 5.12 the relative h-cobordism theorem implies that if N_F is 1-connected and N_F is an h-cobordism then f extends to an orientation reversing diffeomorphism on F.

Thus the situation is very similar to the case of orientation preserving diffeomorphisms. The modifications of N_F below the middle dimension and in the complete odd-dimensional case can be performed into this situation and we obtain the following results.

If M is odd-dimensional then an orientation reversing diffeomorphism f on M is null-bordant if (M,f) has an admissible manifold. If in the even-dim case (M,f) has an admissible manifold then there exists an admissible manifold N_F with the properties of Proposition 6.11.

To complete the even-dimensional case - up to the existence of an admissible manifold - we have to go through Proposition 8.1 , 8.3 and 8.4 and to check the conditions necessary to extend the arguments.

There is no problem with Proposition 8.1 which does not use the fact that the diffeomorphism preserves the orientation. But it is clear that we have to replace one condition in Proposition 8.3 and one in Proposition 8.4 by appropriate conditions for orientation reversing diffeomorphisms. In Proposition 8.4 we had the condition $\tau(N) = \varphi(M,f)$ from which we conclude that $\tau(N_F) = 0$ and we will work with a similar condition in our situation.

To formulate Proposition 8.3 for orientation reversing diffeomorphisms we must define the isometric structure of an orientation reversing diffeomorphism. As such a diffeomorphism doesn't preserve the intersection form but changes the sign, we introduce the following Witt group:

Definition 9.10: Let $W^-_\epsilon(\mathbb{Z};\mathbb{Z})$ be the Witt group of ϵ-symmetric unimodular bilinear forms with anti-isometry h, where h is an anti-isometry if $s(v,w) = -s(h(v), h(w))$.

Then we have an obvious bordism invariant

$$I: \Delta^-_{2m} \longrightarrow W^-_{(-1)^n}(\mathbb{Z};\mathbb{Z})$$

again denoted as isometric structure.

The following Proposition will play the role of Proposition 8.3:

Proposition 9.11: Let N_F be a 2k-dimensional admissible manifold for an orientation reversing diffeomorphism (M,f) with the properties of $N'_F{}'$ in Proposition 6.11. If I(M,f) vanishes in $W^-_{(-1)^k}(\mathbb{Z};\mathbb{Z})$ then by a sequence of additions of handles to N_F we can replace N_F by an admissible manifold $N'_F{}'$ with the same properties as in Proposition 8.3.

The proof is the same as for Proppsition 8.3. One needs only a check that the proof carries over to the case of orientation reversing diffeomorphisms. In the proof of Proposition 8.3 the diffeomorphism enters only in the verification of condition 1.). There is one argument which is wrong in the orientation reversing case. Namely we applied there the fact that $j_*(x-f_*x)$ maps to zero in $H_{k-1}(N_F) = H_{k-1}(BSO)$. But as $H^*(BSO;\mathbb{Q})$ is a polynomial ring in the Pontrjagin classes and f^* preserves Pontrjagin classes it follows that $j_*(x-f_*x)$ maps to zero in $H_{k-1}(N_F;\mathbb{Q})$ which is sufficient as condition 1.), is a statement over $\mathbb{Q}$.

The other point in the proof of Proposition 8.3. where f is involved is where we use that if $x \circ y = 0$ then $f_* x \circ f_* y = 0$, but this is true for orientation reversing diffeomorphisms, too.

Thus all the arguments go through showing Proposition 9.11.

Next we translate Proposition 8.4 into our situation. There is one difference between the admissible manifold for orientation preserving and reversing diffeomorphisms. In the first case $\tau(N_F)$ is invarinat under additions and subtractions of handles but in the orientation reversing case this is not true. In this case the signature mod 2 is a bordism invariant and it turns out that this is the right condition in the orientation reversing case.

<u>Proposition 9.12:</u> Let N_F be a 2k-dimensional admissible manifold of an orientation reversing diffeomorphism which fulfils the properties of $N'_F{}'$ in Proposition 9.11 and, for k even, satisfies: $\tau(N_F) = 0 \bmod 2$. Then, by a finite sequence of handle additions one may replace N_F by $N'_F{}'$ with the same properties as in Proposition 8.4.

The proof of Proposition 8.4 again works in our situation if we assume that N_F has vanishing signature. Thus it is sufficient to show that we can change the signature of N_F by an arbitrary even number using a sequence of additions of k-dimensional handles such that the assumptions are still satisfied. The last is certainly fulfilled if we change N_F by boundary connected sum with two copies of the disk bundle of the tangent bundle of S^k where we glue within F_0 and F_1. In our case where F_0 and F_1 have the same orientation this raises the signature by 2. This shows that we can change the signature by an arbitrary even number.

Summarizing Proposition 6.10 and 8.1 - translated into the orientation reserving case - and Proposition 9.11 and 9.12 we obtain the following result.

Theorem 9.13: $m \gtrsim 5$. Let N_F be an m-dim admissible manifold of an orientation reversing diffeomorphism (M,f) and for dim N_F = 0 mod 4 assume that $\tau(N_F)$ is even. Then by a sequence of surgeries of N_F and additions and subtractions of handles we can replace N_F by a relative h-cobordism.

Remark 9.14: The Theorem is also true for $M = \emptyset$ with the provision that an admissible manifold is a manifold whose boundary consists of two (non-empty) copies of a closed manifold F with same orientation. The bordism group of such admissible manifolds is a geometric model for the ordinary oriented bordism group with $\mathbb{Z}_2$-coefficients denoted by $\Omega_m(pt;\mathbb{Z}_2) \cong \pi_{m+1}(MSO \wedge M(\mathbb{Z}_2))$, where $M(\mathbb{Z}_2)$ is the Moore space obtained by adding a 2-cell to S^1 with a degree two map [45]. Thus we obtain as a special case of Theorem 9.13 the following.

Corollary 9.15: An element in $\Omega_m(pt;\mathbb{Z}_2)$ is bordant to an h-cobordism if and only if the signature is even.

We may apply this result to the following problem. We call a manifold reversible if it admits an orientation reversing diffeomorphism. In general its a difficult problem to decide which manifolds are reversible [18] . Our next result gives the answer up to bordism. The result can also be obtained from [46] by showing that the 2-torsion of Ω_* is generated by reversible manifolds (compare the discussion of generators in § 11).

Theorem 9.16: A closed oriented manifold is bordant to a reversible manifold if and only if it has order 2 in Ω_* or equivalently if all Pontrjagin numbers vanish.

Proof: For bordism with $\mathbb{Z}_2$-coefficients one has a universal coefficient Theorem:

$$0 \rightarrow \Omega_m \otimes \mathbb{Z}_2 \rightarrow \Omega_m(pt;\mathbb{Z}_2) \rightarrow \text{2-Tor}\, \Omega_{m-1} \rightarrow 0$$

The boundary operator is obtained by assigning to a manifold whose boundary consists of two copies of a manifold F this manifold F.

So if $[F]$ is 2-torsion in Ω_{m-1} it is the boundary of an element N_F in $\Omega_m(pt;\mathbb{Z}_2)$ and, without loss of generality, we can assume that $\tau(N_F)$ is even. Then we know that we can replace N_F by a sequence of surgeries and additions and subtractions of handles by a bordant manifold $N'_{F'}$, which is an h-cobordism. Thus F' admits an orientation reversing diffeomorphism and F' is bordant to F.

q.e.d.

We now have a complete invariant for the bordism class of an orientation reversing diffeomorphism. In the odd-dimensional case the only obstruction is the existence of an admissible manifold. But we have an obvious invariant for this obstruction. If (M,f) is an orientation reversing diffeomorphism we can consider M x I as an element of $\Omega_{m+1}(pt;\mathbb{Z}_2)$ identifying $M \times \{0\}$ with $M \times \{1\}$ under f so that the two copies of M in $\partial(M \times I)$ have the same orientation. We denote this element in $\Omega_{m+1}(pt;\mathbb{Z}_2)$ by $M_{(f)}$. Now the vanishing of $M_{(f)}$ is by definition equivalent to the existence of an admissible manifold.

For m even the second invariant is the isometric structure I(M,F). From Theorem 9.13 we know that these two invariants give a complete classification for (M,f) in Δ_m^- for $m \geq 3$. For the only additional invariant in Theorem 9.13 is the signature mod 2 of the admissible manifold but we can assume this to be even by adding a complex projective space to N_F.

To finish the computation of Δ_m^- we have to examine the image of the invariant. But Theorem 9.13 shows that the only obstruction for an element in the bordism group $\Omega_{m+1}(pt;\mathbb{Z}_2)$ to be represented by an h-cobordism (this is equivalent to beeing in the image of $\Delta_m^- \longrightarrow \Omega_{m+1}(pt;\mathbb{Z}_2)$) is the signature mod 2. Denoting the kernel of the signature mod 2 by $\hat{\Omega}_{m+1}(pt;\mathbb{Z}_2)$ we get a surjective map

$$\Delta_m^- \longrightarrow \hat{\Omega}_{m+1}(pt;\mathbb{Z}_2).$$

To realize the isometric structure we distinguish between the case m=2 mod 4 and m=0 mod 4. As in the orientation preserving case it is enough to show the surjectivity for m=2 and m=4. In the case m=2 the result follows from the fact that every oriented surface admits an orientation reversing diffeomorphism. So as every isometry of $H_1(F)$ can be represented by a diffeomorphism the same is true for an anti-isometry. The same argument applies to m=4. As for an anti-isometric structure the signature is zero , it is equivalent to an anti-isometry of $H_2(kP_2\mathbb{C} \# k(-P_2\mathbb{C})\# S^2 \times S^2)$, a manifold which admits an orientation reversing diffeomorphism. This implies:

$$\Delta^-_{2m} \longrightarrow W_{(-1)^m}(\mathbb{Z};\mathbb{Z})$$

is surjective.

Moreover, for m odd and for each $a \in W^-_{(-1)^m}(\mathbb{Z};\mathbb{Z})$ there exists a (M,f) with I(M,f) = a and $M_{(f)} = 0$, for this is true if m = 1 because $\Omega_3(pt;\mathbb{Z}_2) = 0$. This implies that $\Delta^-_{2m+2} \longrightarrow W^-_{(-1)}(\mathbb{Z};\mathbb{Z}) \oplus \Omega_{4m+3}(pt;\mathbb{Z}_2)$ is surjective. For m = 2 we were unable to decide if there exists a (M,f) with $M_{(f)} \neq 0$ in $\Omega_5(pt;\mathbb{Z}_2) \cong \mathbb{Z}_2$ and I(M,f) = 0. If this is true then $\Delta^-_{4m} \longrightarrow W_{(+1)}(\mathbb{Z};\mathbb{Z}) \oplus \Omega_{4m+1}(pt;\mathbb{Z}_2)$ is surjective. Otherwise there is a cokernel $\mathbb{Z}_2$.

Summarizing we obtain the following result.

<u>Theorem 9.17:</u> Let m > 1. There are isomorphisms

$$\Delta^-_{4m-1} \longrightarrow \hat{\Omega}_{4m}(pt;\mathbb{Z}_2)$$

$$\Delta^-_{4m+1} \longrightarrow \Omega_{4m+2}(pt;\mathbb{Z}_2)$$

$$\Delta^-_{4m+2} \longrightarrow W^-_-(\mathbb{Z};\mathbb{Z}) \oplus \Omega_{4m+3}(pt;\mathbb{Z}_2)$$

and we have an injection

$$\Delta^-_{4m} \hookrightarrow W^-_+(\mathbb{Z};\mathbb{Z}) \oplus \Omega_{4m+1}(pt;\mathbb{Z}_2) \text{ with cokernel } \mathbb{Z}_2 \text{ or } \{0\}.$$

Finally we examine the Witt groups $W_\epsilon^-(\mathbb{Z};\mathbb{Z})$. As previously mentioned the group Δ_m^- consists only of 2-torsion for each element is diffeomorphic to its inverse and the same is true for $W_\epsilon^-(\mathbb{Z};\mathbb{Z})$. On the other hand the group $W_\epsilon^-(\mathbb{Z};\mathbb{Z})$ is not finitely generated. This follows from the fact that the elements we have used in § 3 to show that $W_\epsilon(\mathbb{Z};\mathbb{Z})$ has infinite 2-torsion are squares of anti-isometries. Thus we obtain

<u>Proposition 9.18:</u> $W_\epsilon^-(\mathbb{Z};\mathbb{Z}) \cong \mathbb{Z}_2^\infty$.

Another proof is to define an invariant similar to the characteristic polynomial invariant defined in § 2. For this we consider polynomials $F(t)$ over $\mathbb{Z}$ with $F(0) = \pm 1$ and $F(t) = \pm t^d F(-t^{-1})$ where d is the degree of F. Similar as in § 2 we say that F_1 and F_2 are equivalent, if $F_1(t)\cdot F_2(t)$ can be written in the form $\pm t^k f(t)\cdot f(-t^{-1})$ where k is the degree of f and $2k = \text{degree } F_1 + \text{degree } F_2$.

Then for $(V,s,f) \in W_\epsilon^-(\mathbb{Z};\mathbb{Z})$ an easy computation shows that $\det(f-t\cdot \mathrm{Id})$ has the properties of F above and that it's equivalence class is an invariant for $W_\epsilon^-(\mathbb{Z};\mathbb{Z})$. With this invariant it is not difficult to construct infinitely many elements in $W_\epsilon^-(\mathbb{Z};\mathbb{Z})$. A more detailed investigation of $W_\epsilon^-(\mathbb{Z};\mathbb{Z})$ is contained in ([50]).

As mentioned above the case of bordism of diffeomorphisms of non-oriented manifolds is implicitely contained in Quinn's paper [38] . For completeness we state the result here.

Let Δ_n^o be the bordism group of diffeomorphisms of non-oriented manifolds. The main invariant in this case again lives in a Witt group

$W^s(\mathbb{Z}[\mathbb{Z}_2], -)$. Here the involution is defined as follows. If we identify the group ring $\mathbb{Z}[\mathbb{Z}_2]$ with the ring of polynomials in a variable t with the relation $t^2 = t^0$ then - is the involution $n \cdot t^0 + m \cdot t \longmapsto n \cdot t^0 - m \cdot t$. The elements in $W^s(\mathbb{Z}[\mathbb{Z}_2], -)$ are represented by sesquilinear forms over free finitely generated $\mathbb{Z}[\mathbb{Z}_2]$-moduls. For details see [38].

Theorem 9.19 (Quinn [38]): $n \geq 1$. There is an exact sequence

$$0 \longrightarrow \Delta^0_{2n+1} \longrightarrow \mathfrak{N}_{2n+1} \oplus \mathfrak{N}_{2n+2} \xrightarrow{(o,i)} W^s(\mathbb{Z}[\mathbb{Z}_2], -) \longrightarrow$$

$$\Delta^0_{2n+2} \longrightarrow \mathfrak{N}_{2n+2} \oplus \mathfrak{N}_{2n+3} \longrightarrow 0$$

We can split this sequence into the following two short sequences.

Theorem 9.20 (Quinn [38]): For $n \geq 1$ there are exact sequences

a) $0 \rightarrow \Delta^0_{2n+1} \longrightarrow \mathfrak{N}_{2n+1} \oplus \mathfrak{N}_{2n+2} \rightarrow \mathbb{Z}_2 \longrightarrow 0$

where the map into $\mathbb{Z}_2$ is the Euler characteristic of the element in $\mathfrak{N}_{2n+2}$ mod 2.

b) $0 \longrightarrow W^s(\mathbb{Z}[\mathbb{Z}_2], -)/_{i(cl)} \longrightarrow \Delta^0_{2n+2} \rightarrow \mathfrak{N}_{2n+2} \oplus \mathfrak{N}_{2n+3} \longrightarrow 0$

where i is Quinn's invariant for open books on non-orientable manifolds with fundamental group $\mathbb{Z}_2$ and i (cl) the image of this invariant for closed manifolds.

Proof: a) $[M_f] = 0 \Rightarrow M_f = \partial W$ and the mapping torus defines an open book decomposition on ∂W. As W is odd-dimensional this open book decomposition extends to W. By definition this implies that there exists a manifold F with $\partial F = M + M'$ and a diffeomorphism g on F s.t. $g|_M = f$ and $g|_{M'} = Id$. Thus $[M,f] = [M', Id]$ and if in addition $[M] = 0$ then $[M,f] = 0$. This shows the injectivity in the exact sequence.

It remains to show that the only obstruction for the representability of a bordism class by a fibration over S^1 is the Euler characteristic mod 2. But this was shown in [12].

b) The surjectivity on the right side follows from the same result as above. It suffices to show that the kernel of $[M,f] \longmapsto ([M], [M_f])$ is isomorphic to $W^s(\mathbb{Z}[\mathbb{Z}_2], -)/_{i(cl)}$.

For this we define a map j from this kernel to $W^s(\mathbb{Z}[\mathbb{Z}_2], -)/_{i(cl)}$. Without loss of generality we may assume that $w_1(M) \neq 0$. Now if $[M_f] = 0$ then there exists a connected W with $\partial W = M_f$. We can assume that $\pi_1(W) = \mathbb{Z}_2$. Then j(M,f) shall be represented by i(W).

We must show that this is well defined. If $M_f = \partial W'$ with W' connected and $\pi_1(W') = Z_2$, then it is not difficult to show that i(W) and i(W') differ by $i(W \cup_{M_f} W')$.

The injectivity of j can be seen as follows. If $[M] = 0$ and $[M_f] = 0$ and $j(M,f) = 0$ we can assume that M_f bounds a W with $i(W) = 0$. For then there exist W_1, W_2 with $\partial W_1 = M_f$ and $\partial W_2 = \emptyset$ and $i(W_1 = i(W_2)$. But $W_1 \# W_2$ is bordant mod boundary to a manifold W with $\pi_1(W) = \mathbb{Z}_2$ and, by the bordism invariance of i, it follows that $i(W) = 0$. Then the open book decomposition given by the mapping torus on ∂W extends to W and thus $[M,f] = 0$.

Finally the surjectivity of j follows by the realization Theorem of Quinn.

q.e.d.

In conclusion I would like to remark that there are other possible bordism groups of diffeomorphisms which may be computed by similar methods. One is diffeomorphisms of singular manifolds over a topological space X which, for oriented manifolds, is contained in Quinn's work. Furthermore, this could be extended to manifolds with (B,f)-structure.

Another case is orientation reversing diffeomorphisms of (B,f)-manifolds. Our methods extend to this case but we leave the formulation and proof of the results as an exercise.

§ 10 Application to SK-groups

As previously mentioned there is a close connection between the image of the mapping torus invariant $\Delta_n^{(B,f)} \longrightarrow \Omega_{n+1}^{(B,f)}$ and the SK (= cutting and pasting)-groups for (B,f)-manifolds. The SK-theory for (B,f)-manifolds was outlined by G. Barthel [5] and we refer to this for the basic notations and definitions.

Up to a possible $\mathbb{Z}_2$-invariant Barthel has shown that the computation of the SK-groups of (B,f)-manifolds, denoted by $SK_n^{(B,f)}$ is equivalent to the computation of the image of $\Delta_n^{(B,f)} \longrightarrow \Omega_{n+1}^{(B,f)}$ ([5], Theorem 1 and 2). The normal method of computing the SK-groups is to determine this image by constructing appropriate generators of $\Omega_{n+1}^{(B,f)}$. For example Barthel carries this out for weakly complex manifolds. But in general such generators are not known and thus the general computation is still open.

As we have used a completely different approach to compute the image of $\Delta_n^{(B,f)} \longrightarrow \Omega_{n+1}^{(B,f)}$ in § 9 we may avoid the construction of generators. Furthermore we will show that at least for multiplicative (B,f)-structures the $\mathbb{Z}_2$-part mentioned above does not occur completing the computation of $SK_n^{(B,f)}$ in this case.

We first recall the main results which reduce the computation of $SK_n^{(B,f)}$ (up to a $\mathbb{Z}_2$-invariant) to the image of $\Delta_n^{(B,f)} \longrightarrow \Omega_{n+1}^{(B,f)}$. As before we only consider oriented (B,f)-manifolds.

<u>Theorem ([5], Theorem 1 and 2).</u> There are exact sequences

$$0 \longrightarrow I_n^{(B,f)} \longrightarrow SK_n^{(B,f)} \longrightarrow \overline{SK}_n^{(B,f)} \longrightarrow 0 \text{ and}$$

$$0 \longrightarrow (\mathrm{im}\, \Delta_{n-1}^{(B,f)} \longrightarrow \Omega_n^{(B,f)}) \longrightarrow \Omega_n^{(B,f)} \longrightarrow \overline{SK}_n^{(B,f)} \longrightarrow 0$$

where $I_n^{(B,f)}$ is the subgroup generated by S^n with the restriction of the (B,f)-structure on D^{n+1} given by the standard orientation.

$I_n^{(B,f)} \cong \mathbb{Z}$ for n even and $I_n^{(B,f)} = \mathbb{Z}_2$ or $\{0\}$ for n odd.

Lemma 10.1: If (B,f) is a multiplicative structure then $I_{2n+1}^{(B,f)} = 0$.

Proof: We will show that S^{2n+1} vanishes with any (B,f)-structure in $SK_{2n+1}^{(B,f)}$. We do this inductively. It is obvious for S^1 ([5], Lemma 3.i). Next we decompose S^{2n+1} as $S^n \times D^{n+1} \cup D^{n+1} \times S^n$. On the intersection $S^n \times S^n$ we have the product of the trivial (B,f)-structure (given by restriction from D^{n+1}) with itself. For, if we consider it as the boundary of $S^n \times D^{n+1}$, then we see that it has the product structure of some structure on S^n with the trivial structure. If we change the orientation we can consider it as the boundary of $D^{n+1} \times S^n$, so it must be the square of the trivial structure of S^n, where we assume that the orientation is appropriately choosen.

Now, if n is even, the involution $T: S^n \times S^n \longrightarrow S^n \times S^n$, $(x,y) \longmapsto (y,x)$ preserves the square of the trivial structure on S^n. This implies that for n even S^{2n+1} is SK-equivalent to $S^n \times S^{n+1}$ with the product of the trivial (B,f)-structures on it. For n odd we obtain the same result. For then we consider $(1 \times R) \circ T$, where R is reflection on S^n at S^{n-1} which changes the orientation and extends to D^{n+1}. So $S^{2n+1} = S^n \times D^{n+1} \cup_{Id} D^{n+1} \times S^n$ is SK-equivalent to $S^n \times D^{n+1} \cup_{(1 \times R) \circ T} D^{n+1} \times S^n = S^n \times S^{n+1}$. Now by induction it follows that

$S^n \times S^{n+1}$ is zero in $SK_{2n+1}^{(B,f)}$.

q.e.d.

Now we combine this Lemma with the computation of im $(\Delta_{n-1}^{(B,f)} \longrightarrow \Omega_n^{(B,f)})$ in § 9 and Barthel's exact sequences and obtain

Theorem 10.2: Let (B,f) be multiplicative, B 1-connected, all dimensions >4.

Then: $SK_{2n+1}^{(B,f)} = \{0\}$.

$$SK_{4n+2}^{(B,f)} \cong \mathbb{Z}$$

$$SK_{4n}^{(B,f)} \cong \begin{cases} \mathbb{Z} \oplus \mathbb{Z} & \text{if there exists a (B,f)-manifold with non-trivial signature} \\ \mathbb{Z} & \text{otherwise.} \end{cases}$$

The non-trivial summands are detected by Euler-characteristic and signature but the image of these invariants obviously depends on the given (B,f)-structure.

Finally we can apply our results to get some information about SK-groups of diffeomorphisms. We are able to define these groups for diffeomorphisms of (B,f)-manifolds but we give the definition here only in the oriented category.

Let $M_1 = N \bigcup_{\varphi} -N'$ and $M_2 = N \bigcup_{\psi} -N'$ be closed oriented manifolds obtained by cutting along ∂N and glueing with orientation preserving diffeomorphisms φ and ψ. Consider orientation preserving diffeomorphisms f_i on M_i which preserve N and N' and suppose that $f_1|_N$ is isotopic to $f_2|_N$ and the same for N'. Then we say that (M_2,f_2) is obtained from (M_1,f_1) by a SK-move.

We call two diffeomorphisms (M_1,f_1) and (M_2,f_2) SK-equivalent, if there exists a diffeomorphism (M,f) such that $(M_2,f_2) + (M,f)$ can be obtained from $(M_1,f_1) + (M,f)$ by a finite number of SK-moves. The Grothendieck group of the n-dimensional equivalence class under disjoint union is denoted by $SK_{(\mathbb{Z},n)}$. If we further devide out the zero-bordant diffeomorphisms we denote the resulting group by $\overline{SK}_{(\mathbb{Z},n)}$.

The kernel of the projection map $\Delta_n \longrightarrow \overline{SK}_{(\mathbb{Z},n)}$ can be described as follows.

<u>Proposition 10.3:</u> Let $F_{(\mathbb{Z},n)} \subset \Delta_n$ be the subgroup of all elements which can be represented by a differentiable fibre bundle over S^1 commuting with the diffeomorphism then the sequence

$$0 \longrightarrow F_{(\mathbb{Z},n)} \longrightarrow \Delta_n \longrightarrow \overline{SK}_{(\mathbb{Z},n)} \longrightarrow 0 \quad \text{is exact.}$$

<u>Proof:</u> Translate the proof in ([21], p.16.ff) to this setting.

This Proposition and the computation of $F_{(\mathbb{Z},n)}$ is the first step towards a computation of the SK-groups. The next, and in our case, more difficult step would be the determination of the kernel of $SK_{(\mathbb{Z},n)} \longrightarrow \overline{SK}_{(\mathbb{Z},n)}$. The corresponding computation for SK_n is based on handle decomposition and the proof could only be generalized to $SK_{(\mathbb{Z},n)}$ only if one had an equivariant handle decomposition. Therefore the computation of this kernel seems difficult and remains an open problem.

We will now complete the first step namely the computation of $\overline{SK}_{(\mathbb{Z},n)}$. From Lemma 3.3 we know that the isometric structure is a SK-invariant. Thus we obtain a homomorphism $I: \overline{SK}_{(\mathbb{Z},2n)} \longrightarrow W_{(-1)^n}(\mathbb{Z},\mathbb{Z})$.

Theorem 10.4: For oriented or stably almost complex manifolds with $\mathbb{Z}$-actions of dimension >3 we have:

$$\overline{SK}_{(\mathbb{Z},2n+1)} = \{0\} \text{ and}$$

$$I: \overline{SK}_{(\mathbb{Z},2n)} \longrightarrow W_{(-1)^n}(\mathbb{Z};\mathbb{Z}) \text{ is an isomorphism.}$$

Proof: Because I is surjective we need only to show its injectivity. We do this in the oriented case, the proof in the stably almost complex case is exactly the same. Given (M,f) with $I(M,f) = 0$, we have that sign $M = 0$. Hence M is equivalent to zero in $\overline{SK}_n$ and so (M,Id) is equivalent to zero in $\overline{SK}_{(\mathbb{Z},n)}$. Thus we may assume that $[M] = 0$ in Ω_n. By Proposition 10.3 it suffices to show that, if $I(M,f) = 0$ and $[M] = 0$ in Ω_n then $[M,f]$ is contained in $F_{(\mathbb{Z},n)}$. For this we consider $[M_f] \in \Omega_{n+1}$. Next we use the result of Alexander and Kahn that the only obstructions for representing a manifold in its bordism class by a fibration over the torus is the signature and in dim 4k+1 the de Rham invariant [3]. Since $I(M,f)=0$ implies that the de Rham invariant vanishes (Lemma 4.4) we can find a fibration N over $S^1 \times S^1$ such that N is bordant to M_f. N is given by a pair of diffeomorphisms g,h on the fibre F together with an isotopy H_t between Id and the commutator $[g,h]$. The diffeomorphisms g and h classify

$N|_{S^1 \vee S^1}$ and the isotopy gives the extension of this fibration to $S^1 \times S^1$. Now it's easy to see that this isotopy H_t can be used to extend h to a diffeomorphism $\tilde{h}$ on the mapping torus F_g such that the mapping torus $(F_g)_{\tilde{h}}$ is just N.

But $[F_g, \tilde{h}]$ is contained in $F_{(\mathbb{Z},n)}$. From the Wang sequence of F_g one can see that $I(F_g, \tilde{h}) = 0$. Thus $[F_g, \tilde{h}] - [F_g, Id]$ is equal to $[M,f]$ in Δ_n. On the other hand as we mentioned already above $[F_g, Id]$ is zero in $\overline{SK}_{(\mathbb{Z},n)}$ and so $[M,f]$ is zero in $\overline{SK}_{(\mathbb{Z}\, n)}$.

q.e.d.

§ 11 Miscellaneous results: Ring structure, generators, relation to the inertia group.

A) Ring structure. $\Delta_* = \oplus \Delta_m$ has the structure of a graded ring by cartesian product. Furthermore $W_+(\mathbb{Z};\mathbb{Z}) \oplus W_-(\mathbb{Z};\mathbb{Z})$ has a ring structure by tensor product.

We introduce the graded ring W_* with $W_m = \begin{cases} 0 & m \text{ odd} \\ W_+(\mathbb{Z};\mathbb{Z}) & m = 0 \bmod 4 \\ W_-(\mathbb{Z};\mathbb{Z}) & m = 2 \bmod 4 \end{cases}$

Finally we introduce a graded ring 0_* with $0_m = \Omega_m \oplus \Omega_{m+1}$ and multiplication given by

$$([M],[N]) \cdot ([M'],[N']) := ([M\times M'], [M\times N'] + [N\times M'])$$

The ring structure of 0_* can be determined by the ring structure of Ω_* which is well known [46].

Theorem 11.1: The map $\Delta_* \longrightarrow \oplus_m (W_m \times 0_m)$, $[M,f] \longmapsto (I(M,f), [M], [M_f])$ is a ring homomorphism.

Proof: The multiplicativity of I follows from Lemma 3.3. The only non-obvious statement is that $[M\times N_{f\times g}] = [M_f \times N] + [M\times N_g]$.
But $[M\times N_{f\times g}] = [M\times N_{(f\times Id)\cdot(Id\times g)}] = [M\times N_{f\times Id}] + [M\times N_{Id\times g}]$ (by Remark 4.1)$= [M_f\times N] + [M\times N_g]$.

q.e.d.

As this homomorphism is injective for $* \neq 2$ and we can describe its image

(Theorem 5.7), Theorem 11.1 describes the ring structure of the ring $\bigoplus_{m\neq 2}\Delta_m$.

B) Generators. The problem of constructing generators of Δ_m splits into two problems. The first is to construct diffeomorphisms which generate $\Omega_m \oplus \hat{\Omega}_{m+1}$ and the second is to construct diffeomorphisms which generate $W_{\pm}(\mathbb{Z};\mathbb{Z})$ but vanish in $\Omega_m \oplus \hat{\Omega}_{m+1}$. For the second problem and $W_+(\mathbb{Z};\mathbb{Z})$ it is enough to construct generators of $\hat{W}_+^{even}(\mathbb{Z};\mathbb{Z})$ (compare Theorem 9.9). In Proposition 9.7 we recalled a construction of diffeomorphisms on $r(S^n \times S^n)$ for $n > 2$ which generate $W_-(\mathbb{Z};\mathbb{Z})$ for n odd and $\hat{W}_+^{even}(\mathbb{Z};\mathbb{Z})$ for n even and vanish in $\Omega_{2m} \oplus \Omega_{2m+1}$. The case n = 2 is more difficult but can be done in a similar way [47].

So we are reduced to the construction of diffeomorphisms which generate $\Omega_m \oplus \hat{\Omega}_{m+1}$ or as generators of Ω_m are well known ([32], [16]), by taking the identity diffeomorphism on them, we need only find diffeomorphisms whose mapping tori generate $\hat{\Omega}_{m+1}$.

The problem splits into two parts: the torsion free part and the torsion case. For the torsion free case there are examples already in the literature [14]. We will repeat this construction later. A similar construction can be done for a majority of the torsion.

We begin by describing two constructions. Given n orientation reversing diffeomorphisms (M_i, f_i) of oriented manifolds we first construct a new oriented manifold. $\mathbb{Z}^{n-1}$ operates on $M_1 \times \ldots \times M_n$ by $e_i(x_1,\ldots,x_n) = (x_1,\ldots,x_{i-1}, f_i x_i, f_{i+1}x_{i+1}, x_{i+2},\ldots,x_n)$. We define $(M_1,\ldots,M_n)_{(f_1,\ldots,f_n)}$ as the bundle over T^{n-1} given by this operation:

$(M_1,\ldots,f_n)_{(f_1,\ldots,f_n)} := \mathbb{R}^{n-1} \times_{\mathbb{Z}^n} M_i$. The diffeomorphism

$(x_1,\ldots,x_n) \longmapsto (x_1,\ldots,x_{n-1}, f_n(x_n)$ induces a diffeomorphism $\tilde{f}_n$ on

$(M_1,\ldots,M_n)_{(f_1,\ldots,f_n)}$. It follows by construction that

$(M_1,\ldots,M_n)_{(f_1,\ldots,f_n)}$ is the mapping torus of the diffeomorphism $\tilde{f}_{n-1} \times f_n$

operating on $(M_1,\ldots,M_{n-1})_{(f_1,\ldots,f_{n-1})} \times M_n$.

The other construction starts with an S^1-action and ends up with a diffeomorphism on the base space of a fibre bundle with fibre T^n, the n-dimensional torus. Given a S^1-action on M we construct a manifold V(n,M) as follows. T^n operates freely on $(S^3)^n \times M$ by

$$t_i((x_1,y_1),\ldots,(x_n,y_n), a) = \begin{cases} ((t_1x_1,t_1y_1),(x_2,y_2),\ldots,(x_n,y_n),t_1^{-1}a) \text{ for } i=1 \\ ((x_1,y_1),\ldots,(x_{i-2},y_{i-2}),(x_{i-2},t_i^{-1}y_{i-1}), \\ (t_ix_i,t_iy_i),(x_{i+1},y_{i+1}),\ldots,(x_n,y_n),a) \text{ for } i>1 \end{cases}$$

$V(n,M) := (S^3)^n \times M/_{T^n}$. There is a S^1-action α on V(n,M) induced by $t(x_1y_1),\ldots,(x_n,y_n), a) = ((x_1,y_1),\ldots,(x_{n-1},y_{n-1}), (x_n,t\,y_n), a)$.
Again by construction $V(n,M) = S^3 \times_{S^1} V(n-1,M)$.

For a S^1-action α on M given by a homomorphism $\alpha: S^1 \longmapsto \mathrm{Diff}_+(M)$ we denote the corresponding diffeomorphism on $S^1 \times M$ by $f_\alpha : (t,x) \longmapsto (t,\alpha(t)\cdot x)$. It was first remarked by Burdick that the mapping torus of f_α is bordant to $S^3 \times_{S^1} M$ [12].

We will see that the constructions above applied to some special situations lead to a set of generators of Tor Ω_*. The special situation comes from the Dold manifolds. The Dold manifold $P(m,n)$ is the quotient of the free involution on $S^m \times P_n\mathbb{C}$ mapping $(x,z) \longmapsto (-x,\bar{z})$. Let $f(m,n)$ be the involution on $P(m,n)$ induced by reflection in an equator of S^m. For m odd and n even $P(m,n)$ is oriented and $f(m,n)$ is orientation reversing. On $P(1,2n)$ the left multiplication of S^1 on S^1 induces a S^1-action.

<u>Proposition 11.2:</u> Tor Ω_* is generated as a ring by the mapping tori of the following diffeomorphisms

1.) $\left[(P(m_1,n_1),\dots, P(m_k,n_k))_{(f(m_1,n_1),\dots, f(m_k,n_k))} \times P(m,n), \tilde{f}(m_k,n_k) \times f(m,n) \right]$

2.) $\left[V(m-1, P(1,n)), f_\alpha \right]$

3.) conjugation of all coordinates of $P_{2n}\mathbb{C}$.

We can restrict all (m_i,n_i), (m,n) to be of the form $(2^r-1, 2^r \cdot s)$.

<u>Proof:</u> According to Wall Tor Ω_* is generated as a ring by classes $\partial_3 (\prod_{i=1}^{k} M_{i(f_i)})$ where (M_i,f_i) are the diffeomorphisms on the Dold manifolds described above and the indices of the Dold manifolds are as in Proposition 11.2 and ∂_3 is the dual of the first Stiefel-Whitney class [46].

In particular if M_f is the mapping torus of an orientation reversing diffeomorphism of an oriented manifold M then $\partial_3(M_f) = [M]$. This implies, if $k = 1$, then we have $\partial_3(M_f) = [M]$, where M is a Dold manifold $P(2m+1, 2n)$.

If $m = 0$ then $P(1, 2n) = (S^1 \times P_{2n}\mathbb{C})/_T$ and this is equal to the mapping torus of the conjugation on $P_{2n}\mathbb{C}$, also contained in our list.

If $m > 0$ then $P(2m+1, 2n) = (S^{2m+1} \times P_{2n}\mathbb{C})/_T$ is the total space of a fibration over $P_m\mathbb{C}$ which can be described as follows. $P(2m+1), 2n) = S^{2m+1} \times_{S^1} (S^1 \times P_{2n}\mathbb{C})/_T =$
$= S^{2m+1} \times_{S^1} P(1, 2n)$ where the S^1-action on $(S^1 \times P_n\mathbb{C})/_T$ is as described above.

Next we need the following relation for fiberings over projective spaces which was used first by Jänich for the SK-relation but which is also true up to bordism. Let S^1 operate on M. Then in Ω_* we have the equation:

$$\left[S^{2m+1} \times_{S^1} M \right] = \left[P_m\mathbb{C} \times M \right] + \left[S^{2m-1} \times_{S^1} (S^3 \times_{S^1} M) \right]$$

where S^1 operates on $S^3 \times_{S^1} M$ by $t\left[(x,y), a\right] = \left[(x,ty), a\right]$. A bordism between these manifolds is obtained by the disjoint union of $S^{2m+1} \times_{S^1} M \times I$ and $-P_m\mathbb{C} \times M \times I$ and by identification of $D^{2m} \times M$ sitting in $S^{2m+1} \times_{S^1} M \times \{1\}$ and in $-P_m\mathbb{C} \times M \times \{1\}$. The boundary of this manifold consists of $S^{2m+1} \times_{S^1} M$, $-P_m\mathbb{C} \times M$ and a fibration over $P_m\mathbb{C} \# -P_m\mathbb{C}$ with fibre M. One easily checks that the total space of this fibration is $S^{2m-1} \times_{S^1} (S^3 \times_{S^1} M)$ using the fibration of $P_m\mathbb{C} \# -P_m\mathbb{C}$ over $P_{m-1}\mathbb{C}$ with fibre S^2.

Applying this equation inductively we see that $S^{2m+1} \times_{S^1} (S^1 \times P_{2n}\mathbb{C})/_T$ is, up to decomposable manifolds, bordant to the total space of a fibre bundle over S^2 classified by the S^1-action α on $V(m-1\ , P(1,n)$. But according to the result of Burdick [12] such a total space is bordant to the mapping torus f_α.

To finish the proof we have to show that $\partial_3(\prod_{i=1}^{k} M_{i(f_i)})$ for $k > 1$ is contained in our list. We use the following connection between mapping tori.

We have a diffeomorphism between $M_f x N_g$ and $((MxN)_{(fxg)})_{\tilde{g}}$ induced by the map $(M x \mathbb{R}) \times (N x \mathbb{R}) \longrightarrow (M x N x \mathbb{R}) \times \mathbb{R}$ which maps $((x,t), (y,s))$ to $((x,y,t), s-t)$. But we are then finished for this by induction implies that the product $\prod_{i=1}^{k} M_{i(f_i)}$ with $k > 1$ is equal to $((M_1,\dots,M_k)_{(f_1,\dots,f_k)})_{\tilde{f}_k}$. As mentioned before this implies that $\partial_3 \prod_{i=1}^{k} (M_i)_{(f_i)} = (M_1,\dots,M_k)_{(f_1,\dots,f_k)}$. But $k > 1$ and the inductive formula for $(M_1,\dots,M_k)_{(f_1,\dots,f_k)}$ imply that $(M_1,\dots,M_k)_{(f_1,\dots,f_k)}$ is the mapping torus $((M_1,\dots,M_{k-1})_{(f_1,\dots,f_{k-1})} \times M_k)_{\tilde{f}_{k-1} \times f_k}$.

q.e.d.

To finish the description of a set of generators for Ω_* as mapping tori we must find such generators for the free part of it. This is contained in Conner's paper [14]. We repeat his description.

Consider for non-negative integers n and k the free action of the (n+1)-dimensional torus T^{n+1} on $S^{2k+1} \times (S^3)^n$ by

$$(t_1,\dots,t_{n+1}) \cdot [(\lambda_1,\dots,\lambda_{k+1}), (z_2,w_2),\dots, (z_{n+1}, w_{n+1})] =$$

$$[(t_1 t_2^{-1} \lambda_1, t_1 \lambda_2,\dots,t_1 \lambda_{k+1}), (t_2 t_3^{-1} z_2, t_2 \cdot w_2),\dots,(t_n \cdot t_{n+1}^{-1} z_n, t_n \cdot w_n),$$

$$(t_{n+1} z_{n+1}, t_{n+1} w_{n+1})].$$ The orbit manifold is denoted by $V(n,k)$. Now there is a S^1-action α on $V(n,k)$ induced by left multiplication of S^1 on z_{n+1}.

Theorem 11.3 (Conner [14]): Let M_i be a set of ring generators of Ω_*. Consider the subgroup of Δ_* multiplicatively generated by $[S^1 \times V(n,k), f_\alpha]$ and by $[M_i, Id]$. Then modulo an element of order 2 every element of $\hat{\Omega}_*$ is the mapping torus of a diffeomorphism contained in this subgroup.

This completes the list of generators of $\bigoplus_{n\neq 2} \Delta_n$ and we summarize the result as follows.

Theorem 11.4: The diffeomorphisms contained in the following list generate $\bigoplus_{n\neq 2} \Delta_n$ as a ring.

1.) $(r(S^n \times S^n), g)$ where g is as constructed in the proof of Proposition 9.7 and $n > 1$

2.) (M_i, Id) where the M_i form a set of generators of Ω_*

3.) the diffeomorphisms constructed for Proposition 11.2

4.) $(S^1 \times V(n,k), f_\alpha)$.

We want to end our discussion of generators with the following remark. The diffeomorphisms contained in 1.) and 2.) of our list are diffeomorphisms on 1-connected manifolds if we choose M_i 1-connected. But the diffeomorphisms contained in 3.) and 4.) do not operate on 1-connected manifolds. One could ask the question whether Δ_m for $m\neq 2$ is generated by 1-connected manifolds and whether two bordant diffeomorphisms are bordant by a diffeomorphism on 1-connected manifolds. Both statements are true. This follows from our computation of Δ_m. In fact all manifolds we used to bound diffeomorphism were replaced by 1-connected ones as first step, so the answer to the second question is yes. For the first question it suffices to show that every element of $\widehat{\Omega}_m$ is the mapping torus of a diffeomorphism on a 1-connected manifold, but this follows from the proof of Theorem 5.5.

Remark 11.5: The bordism group of diffeomorphisms on 1-connected manifolds modulo those bounding diffeomorphisms on 1-connected manifolds is equal to Δ_m for $m > 2$.

C) Inertia group. There is a connection between diffeomorphisms on spheres and the inertia group of a manifold which was first used by Winkelnkemper to show that all diffeomorphisms on spheres bound [51]. The inertia group $I(M)$ of a n-dimensional manifold M consists of all homotopy spheres Σ such that $M \# \Sigma \cong M$.

The connection between $I(M)$ and Diff (S^{n-1}) is given by the following Lemma.

Lemma 11.6 ([52], [26], Lemma 21): If f is a diffeomorphism of S^{n-1} let $\Sigma = D^n \cup_f D^n$ be the corresponding homotopy sphere.
Then $\Sigma \in I(M) \Longleftrightarrow$ f extends to a diffeomorphism on $M - \mathring{D}^n$.

Now, it is not difficult to see that the mapping torus of a diffeomorphism f on S^{n-1} bounds. We will prove a more general statement below. This implies that every diffeomorphism on S^{n-1} bounds and by Lemma 11.6: For each homotopy sphere Σ^n $(n \neq 3,4)$ there exists a manifold M such that $\Sigma \in I(M)$. As there are only finitely many homotopy spheres this implies that for $n \neq 3,4$ there exists a manifold M s.t. $I(M) = \Theta_n$.

Corollary 11.7: For all $n \neq 3,4$ there exists a n-dimensional manifold M with $I(M) = \theta_n$, the group of homotopy spheres.

As mentioned above, this result was first obtained by Winkelnkemper as a consequence of his Equator Theorem ([51]) and used by him to show that every diffeomorphism on S^n bounds.

We conclude with the Lemma announced above.

Lemma 11.8: If all rational Pontrjagin classes and all Stiefel Whitney classes of M vanish then the mapping torus of every diffeomorphism on M bounds.

Proof: We have to show that all characteristic numbers of M_f vanish. For this we consider the Wang sequence with coefficients $\mathbb{Q}$ in the case of the Pontrjagin classes and $\mathbb{Z}_2$ for the Stiefel-Whitney classes [44].

$$H^{k-1}(M) \xrightarrow{\delta} H^k(M_f) \longrightarrow H^k(M)$$

As the characteristic classes of M vanish all characteristic classes of M_f lie in im δ. On the other hand $\delta(x) \cup \delta(y) = 0$. Thus all characteristic numbers of M_f vanish except perhaps the top Pontrjagin class and the top Stiefel-Whitney class. But if all other characteristic numbers vanish they are determined by the signature [20] and Euler characteristic of M_f which vanish.

q.e.d.

References

[1] J.F. Adams: On the group J(X) - IV, Top. 5 (1966), 21-71

[2] J. Alexander, G. Hamrick and J. Vick: Linking forms and maps of odd prime order Trans. AMS 221 (1976), 169-185

[3] J.C. Alexander and S.M. Kahn: Characteristic number obstructions to fibering oriented and complex manifolds over surfaces, Top. 19 (1980), 265-282

[4] M.F. Atiyah and I.M. Singer: The index of elliptic operators III, Ann. of Math. 87 (1968), 546-604

[5] G. Barthel: Cutting and pasting of (B,f)-manifolds, Appendix in [21] , Publish or Perish Math.Lect.Ser. 1 (1973), 62-70

[6] F. Bonahon: Cobordisme des difféomorphismes de surfaces, C.R. Acad.Sci. Paris 290 série A (1980), 765-767

[7] T. Bröcker and K. Jänich: Einführung in die Differentialtopologie, Springer Verlag (1973)

[8] W. Browder: Surgery and the theory of differentiable transformation groups, Preceedings of the conf. on transformation groups, Ed. Mostert,, Springer Verlag (1968)

[9] W. Browder: Diffeomorphisms of 1-connected manifolds, Trans. A.M.S. 128 (1967), 155-163

[10] W. Browder: Surgery on simply-connected manifolds, Springer Verlag (1972)

[11] W. Browder and J. Levine: Fibering manifolds over a circle, Comm.Math. Helv. 40 (1965/66), 153-160

[12] P.O. Burdick: Orientable manifolds fibred over the circle, Proc.A.M.S. 17 (1966), 449-452

[13] P.E. Conner and E.E. Floyd: Differentiable periodic maps, Springer Verlag (1964)

[14] P.E. Conner: The bordism class of a bundle space, Mich. Math.J. 14 (1967), 289-303

[15] P.E. Conner and E.E. Floyd: Fibering within a bordism class, Mich. Math.J. 12 (1965), 33-47

[16] A. Dold: Erzeugende der Thomschen Algebra $\mathfrak{N}$, Math.Z. 65 (1956), 25-36

[17] A.L. Edmonds and J.H. Ewing: Remarks on the cobordism group of surface diffeo morphisms, Math.Ann. 259 (1982), 497-504

[18] D. Fried: Some irreversible three-manifolds, preprint

[19] P.A. Griffith and J. Morgan: Rational homotopy theory and differentiable forms, Birkhäuser Verlag (1981)

[20] F. Hirzebruch: Topological methods in algebraic geometry, Springer Verlag (1966)

[21] U. Karras, M. Kreck, W. Neumann and E. Ossa: Cutting and pasting of manifolds; S.K.-groups, Publish or Perish Math.Lect. Ser. 1 (1973)

[22] M. Kervaire: Knot cobordism in codimension two, Manifolds - Amsterdam, Springer L.N. 197 (1971), 83-105

[23] M. Kervaire and J. Milnor: Groups of homotopy spheres I, Ann. of Math. 77 (1963), 504-537

[24] M. Kreck: Cobordism of odd-dimensional diffeomorphisms, Top. 15 (1976), 353-361

[25] M. Kreck: Bordism of diffeomorphisms, Bull.A.M.S. 82 (1976), 759-761

[26] M. Kreck: Bordismusgruppen von Diffeomorphismen, Habilitationsschrift, Bonn (1976)

[27] R. Lashof: Poincaré duality and cobordism, Trans. A.M.S. 109 (1963), 257-277

[28] J. Levine: Knot cobordism in codimension two, Comm.Math. Helv. 44 (1969), 229-244

[29] G. Lusztig, J. Milnor and F.P. Peterson: Semi-characteristics and cobordism, Top. 8 (1969), 357-360

[30] S. López de Medrano: Cobordism of diffeomorphisms of (k-1)-connected 2k-manifolds, second conference on compact transformation groups, Amherst, Springer L.N.298 (1972), 217-227

[31] P. Melvin: Bordism of diffeomorphisms, Top. 18 (1979), 173-175

[32] J. Milnor and J. Stasheff: Characteristic classes, Ann. of Math. Studies, 76 (1974)

[33] J. Milnor and D. Husemoller: Symmetric bilinear forms, Springer Verlag (1973)

[34] W. Neumann: Fibering over the circle within a bordism class, Math.Ann. 192 (1971), 191-192

[35] W.D. Neumann: Equivariant Witt rings, Bonner Math.Schr. 100 (1977)

[36] W.D. Neumann: Signature related invariants of manifolds-I. Monodromy and γ-invariants, Top. 18 (1979), 147-172

[37] F. Quinn: The stable topology of 4-manifolds, Top. and its appl. 15 (1983), 71-77

[38] F. Quinn: Open book decompositions, and the bordism of automorphisms, Top. 18 (1979), 55-73

[39] H. Samelson: On the Milnor-Spanier and Atiyah duality theorems, Mich.Math.J. 16 (1969), 1-2

[40] M. Scharlemann: The subgroup of Δ_2 generated by automorphisms of tori, Math.Ann. 251 (1980), 263-268

[41] S. Smale: On the structure of manifolds, Am.J.Math. 84 (1962), 387-399

[42] N.W. Stoltzfus: Unraveling the integral knot concordance group, Mem. A.M.S. 192 (1977)

[43] R.E. Stong: Notes on cobordism theory, Princeton Univ.Press (1968)

[44] E.H. Spanier: Algebraic topology, Mc Graw-Hill (1966)

[45] D. Sullivan: Geometric periodicity and the invariants of manifolds, in Springer L.N. 197 (1971)

[46] C.T.C. Wall: Determination of the cobordism ring, Ann. of Math. 72 (1960), 292-311

[47] C.T.C. Wall: Diffeomorphisms of 4-manifolds, J. London MS. 39 (1964), 131-140

[48] C.T.C. Wall: Classification problems in differential topology II: Diffeomorphisms of handlebodies, Top. 2 (1963), 263-272

[49] C.T.C. Wall: Surgery on compact manifolds, Academic Press (1970)

[50] M. Warshauer: The Witt group of degree k maps and asymmetric inner product spaces, Springer L.N. 914 (1982)

[51] H.E. Winkelnkemper: On equators of manifolds and the action of θ^n, Thesis, Princeton (1970)

[52] H.E.Winkelnkemper: On the action of θ^n I, Trans.A.M.S. 206 (1975), 339-346

Appendix

by Neal W. Stoltzfus

The algebraic relationship between Quinn's Invariant for Open Book Decomposition Bordism and the isometric structure

Dedicated to the memory of my grandparents: Jake and Mable Weiler

Table of Contents:

0) Introduction

The study of diffeomorphisms in the category of smooth manifolds, that is, the automorphism category, has been of great interest since the introduction of the concept of manifold. The investigation of the symmetries or group actions on a manifold has been a major theme and testing ground for results. Utilization of the techniques of bordism in this research was initiated by P.E. Conner and E.E. Floyd in their study of the bordism theory of group actions of groups of finite order. The specific question concerning the nature of the bordism group of arbitrary diffeomorphisms on arbitrary manifolds is due to William Browder who first posed it as a thesis problem to several students. The first results on this general question was the thesis of Elmar Winkelnkemper [W] who demonstrated that any (orientation preserving) diffeomorphism of an arbitrary homotopy sphere extends to a diffeomorphism of some oriented smooth manifold bounding the homotopy sphere. Later he introduced the idea of an open book decomposition of a manifold as a useful technique in this study, generalizing a result on J. Alexander on the structure of three-manifolds. The next advance on this question was the paper of Santiago Lopez de Medrano [LdM], in 1971, in which he demonstrated that the isometric structure of a diffeomorphism on an even dimensional manifold was a strong invariant taking values in an infinitely generated Witt group. The development of this idea followed the lines of a similar computation made earlier by J. Levine [L2] for knot concordance. M. Kreck then obtained the complete solution in 1975 using the isometric structure invariant together with the bordism classes of the underlying manifols and the mapping torus. However, any good problem deserves at least two solutions and F. Quinn provided a second computation in the setting of open book decomppsition bordism utilizing a Witt group of (not necessarily ε-symmetric) unimodular bilinear forms. While the isometric structure takes values in a Witt group which depends on the symmetry type of the intersection pairing (in particular on the parity of the middle dimension), Quinn's invariant has no such parity relationship. In this appendix we develop the explicit relationship between the two invariants. The basic idea is a bilinearization construction which extends a well-known relationship in the theory of high-dimensional fibred knots.

The author of this appendix wishes to thank the National Science Foundation for partial support during the work on this project, the University of Mainz for its hospitality during a stimulating visit when the first rough edges were worked off this relationship, Le Cours de Troisieme Cycle au l'Universite de Geneve for the privilege of given a series of lectures which first piqued my interest in this question and finally to the author of the monograph to which this is an appendix for rekindling my interest in the question of the precise relationship of the isometric structure to Quinn's invariant.

1) The Geometric Context

Let $\Delta_n(X)$ denote the bordism group of orientation preserving diffeomorphisms of n-dimensional oriented manifolds (in the oriented smooth category) over a topological space X. The group consists of triples (P;f,h) where f is a diffeomorphism of P and h is a continuous map to X such that h is homotopic to $_h$ f. Let $\Omega_n^{SO}(X;Z,1)$ be Quinn's group of (n+1)-dimensional book decomposition bordism of a topological space X (usually a point $*$) [Q, p.57]. This group consists of four tuples (P,f,h,H) where

a) P is a compact smooth oriented n-dimensional manifold

b) $h:P \longrightarrow X$ is a continuous map

c) $f:P \longrightarrow P$ is an orientation preserving diffeomorphism which is the identity on the boundary of M

d) H is a homotopy between h and h∘f which is constant on the boundary. The relative "open book mapping torus" t(P,f) is the manifold $P \times_f S^1 \cup \partial P \times D^2$. The homotopy X induces a continuous map from t(P,f) to X. Next, we give the details of the geometrical relationship between the two viewpoints:

<u>Proposition 1:</u> $\Delta_n(X)$ is isometric to $\Omega_n^{SO}(X;Z,1) \oplus \Omega_n^{SO}(X)$ under the homomorphism given by $b(P,f,h) = ([P,f,h,H];[P]), [P]$ the bordism class of the underlying manifold P.

<u>Proof:</u> We will give the proof only in the case that X is a point, the general case following trivially by extension. If (P,f) is zero in the first component then the mapping torus P_f bounds a manifold M with a book decomposition which has page W contained in M, $\partial W = P \cup N$ and there is an automorphism $h:W \longrightarrow W$ satisfying $h|_P = f$ and $h|_N = Id$. Therefore (P,f) is concordant in Δ_n to the (N,Id). Hence, if P bounds, N does also with the identity extending easily to the bounding manifold. Thus we see that b is injective. Surjectivity follows from the identification of a splitting map $s((P,\partial P;h),N) = (D(P) \cup N; (F \cup Id|_P) \cup Id|_N)$ where $D(P) = P \cup P$, the double of P (which always bounds P x I (interpreted appropriately as a manifold with corners at the boundary of a two-sided collar on ∂P)) completing the proof that b∘s is the identity on the second factor. To observe the same on the first factor, we use the product bordism $(D(P) \cup N) \times I$ introducing a corner at ∂P in the "one" end and note that the map on the remainder of $D(P) \cup N$ is the identity. The result is a valid bordism in Quinn's group to $(P, \partial P;h)$.

2) An appropriate Setting for Quinn's Invariant

We now define a group which is the appropriate domain for the invariant of F. Quinn. We will denote this group by $BB_n^{SO}(X;Z,1)$ (BB for bounded book bordism) and it consists of quintuples (M,F;P,f,h,H) where M is a manifold bounding the mapping torus determined by (P,f,h,H) and F is a continuous map to X extending the given one on the mapping torus of f. The bordism relation is defined as follows: the boundary of M, the mapping torus of (P,f) is permitted to vary by a bordism of the open book decomposition over X to another which bounds M', then the union of M, the bordism of the boundary and M' is required to bound over X inducing a bordism of M to M' with corners. Quinn's invariant i is then well defined on the group BB and takes values in A(Z) (our notation for the Witt group of asymmetric bilinear forms.) Furthermore, we can extend Quinn's main result (Theorem 3.2) as follows:

<u>Proposition 2:</u> There is an exact sequence:

$$0 \to \Omega_{2n+1}^{SO}(X;Z,1) \to \Omega_{2n+2}^{SO}(X) \to BB_{2n}^{SO}(X;Z,1) \to \Delta_{2n} \to \Omega_{2n}^{SO}(X) \oplus \Omega_{2n+1}^{SO}(X) \to 0$$

<u>Proof:</u> This exact sequence is obtained from the fundamental exact sequence of F. Quinn [Q, Thm. 3.2] by the addition of the identical group $\Omega_{2n}^{SO}(X)$, to the last two terms in the exact sequence and then making the appropriate identifications via Prop. 1 and the isomorphism $A(Z[\pi_1(X)]) = BB_{2n}^{SO}(X;Z,1)$ which follows from Quinn's Theorem 1.1 (2 and 3). Next we describe the maps in the exact sequence. The first non-trivial map is the relative mapping torus of the book decomposition. The second map takes a manifold over X and views it as the manifold bounding the empty book decomposition. The next

map takes the monodromy on the page, (P,h), doubles P and extends by the identity, $(D(P),h|_P \cup Id.)$ (This follows from the fact that the map from $BB^{SO}_{2n}(X;Z,1)$ to Δ_{2n} factors thru the map from BB to $\Omega^{SO}_{2n}(X,Z,1)$ given by first taking the book decomposition on the boundary and then following by the explicit isomorphism of Prop. 1 which is the stated doubling construction on this component.) The final map takes a diffeomorphism (P,f) to the bordism class of the underlying manifold and the mapping torus, respectively. The purpose of this appendix is to disclose the exact relation between the above exact sequence and the following algebraic exact sequence of Witt groups. The two are related by a commutative diagram involving the invariant of F. Quinn and the isometric structure of a diffeomorphism.

Theorem 1: The following diagram is commutative with exact rows, giving an explicit relationship between the geometric exact sequence of Prop. 2 and an algebraic exact sequence of Witt groups. Furthermore, it formulates the exact relationship between the invariant of Quinn and the isometric structure

$$\begin{array}{ccccccccccc} 0 \rightarrow \Omega^{SO}_{2n+1}(X;Z,1) & \rightarrow & \Omega^{SO}_{2n+2}(X) & \rightarrow & BB^{SO}_{2n}(X;Z,1) & \rightarrow & \Delta_{2n} & \rightarrow & \Omega^{SO}_{2n}(X) & \oplus & \Omega^{SO}_{2n+1}(X) \rightarrow 0 \\ & & \downarrow S & & \downarrow i & & \downarrow I & & \downarrow S & & \downarrow dR \\ & 0 \longrightarrow & W^{-e}(Z) & \longrightarrow & A(Z) & \rightarrow & W^{e}(Z,C_\infty) & \rightarrow & W^{e}(Z) & + & \mathrm{Coker}\, \partial_{-e} \rightarrow 0 \end{array}$$

The maps in the exact sequence can be described as follows: The first map takes a (-e)-symmetric bilinear form and forgets its symmetry. The next map is a "bilinearization" map which will be denoted bil and described in detail in Section 7. The last map forgets the isometry of the form and, again by an explicit construction in Section 7, a second invariant is constructed taking its values in the cokernel of the boundary map ∂_{-e} in the localization sequence for Witt groups.

Using the well-known computations of the Witt groups of integral e-symmetric bilinear forms and the computation of the integral localization sequence the algebraic exact sequences reduce to the following:

$$e = +1 \quad 0 \longrightarrow 0 \longrightarrow A(Z) \longrightarrow W^{1}(Z,C_{\infty}) \rightarrow Z + Z/2 \longrightarrow 0$$

$$e = -1 \quad 0 \longrightarrow Z \longrightarrow A(Z) \longrightarrow W^{-1}(Z,C_{\infty}) \longrightarrow 0$$

In these sequences, the group Z is determined by a signature invariant and the Z/2 by a de Rham invariant (see Section 9).

Computations of the Witt group of isometries have been given in [N] and [St], while the group A(Z) have been computed by Quinn [unpub.] and Warshauer [War].

3) The Standard Duality Diagram

We will make heavy utilization of the following diagram, which will be referred to as the standard (duality) diagram. Let M be a 2k-dimensional manifold with boundary, ∂M, together with an open book decomposition of the boundary with page P. The vertical alignments in the diagram are in perfect duality under the appropriate intersection pairing in M (or ∂M), while the horizontal sequneces are the homology exact sequences of the pair (M,PxI) and the triple, (M, ∂M, ∂M-(PxI)), respectively, together with certain identifications: (For simplicity, all homology groups will have integer coefficients, unless noted otherwise. Secondly, we will consider only the case when X is simply connected. For the general case, local coefficients are needed as explained in Quinn.)

$$H_k(P\times I) \longrightarrow H_k(M) \longrightarrow H_k(M,P\times I) \longrightarrow H_{k-1}(P\times I) \longrightarrow H_{k-1}(M) \longrightarrow H_{k-1}(M,P\times I)$$

$$\begin{array}{cccccc} H_k(\partial M, M-P\times I) \leftarrow H_k(M,\partial M) \leftarrow & H_k(M,P\times I) \leftarrow & H_k(P,\partial P) \leftarrow & H_{k+1}(M,\partial M) \leftarrow & H_{k+1}(M,P\times I) \\ & \cong & \cong & & \cong \\ & H_k(M,\partial M-P\times I) & H_k(\partial M,\partial M-P\times I) & & H_{k+1}(M,\partial M-P\times I) \end{array}$$

The displayed isomorphisms follow from "sliding" in the I coordinate and a "suspension" isomorphism explicated in Quinn [Q] (p.63, bottom and p.59 (4.5) respectively.) We will also need the further following sequence from (4.3) relating the homology of the book and the automorphism:

$$H_k(\partial M) \to H_{k-1}(P,\partial P) \xrightarrow{\ Id\ -\ h_*\ } H_{k-1}(P) \to H_{k-1}(\partial M) \to H_{k-1}(P,\partial P)$$

As commented by Quinn, this follows from the homology exact sequence of the pair (Book = t(h, ∂h=Id), Page = P) and excision and suspension isomorphisms. This is the first indication that a "linking" pairing is involved.

4) The Standard Example: Fibred Knots

The case of fibred knots inside the standard sphere (that is, the complement of the knot is fibred over the circle) can be given a complete and simple analysis relating Quinn's invariant to the standard bilinear linking forms. A good background reference for the following ideas is Alan Durfee's paper [D] : Fibred Knots and Algebraic Singularities (and its further list of references.)

For fibred knots in the standard sphere, we may choose $M = D^{2k}$, $\partial M = S^{2k-1}$, ∂P is the knotted homotopy sphere and P is the Seifert "surface". Quinn's pairing given by deforming the second, say, of two relative cycles in $H_k(M,P)$ so that its boundary is pushed into $\partial M - P$ in the positive normal direction to P in ∂M and then intersecting with the first cycle. But this is the same as the recipe for computing the linking on the Seifert "surface" (Push the second cycle off the surface in the positive normal direction, allow it to bound in the sphere and measure the intersection with the first cycle lying on the "surface".) We have the first instance of a well-known philosophy: "The computation of the intersection of relative cycles (with disjoint boundary) in the interior of a manifold with boundary is equivalent to measuring the linking of the boundaries of the cycles in the boundary manifold." The explicit statement we will use is given in Durfee [D,p.51]:

<u>Proposition 3:</u> Let x and y correspond to bx and by under the boundary isomorphism b: $H_k(D,P\times I) \cong H_{k-1}(P\times I)$. Then Quinn's pairing $E(x,y) = L(bx,by)$, the linking pairing on the Seifert "surface".

<u>Proof:</u> Apply Lemma 2.3 of Durfee [D] , using general relative cycles and noting that bounding in the second factor (as outlined above) reverses Durfee's convention and his remark.

However, we must note that our philosophical remark is not true in complete generality. In fact, there may be closed cycles (redundant, of course) with non-trivial self intersection and secondly, linking on the boundary need not always be defined. Thus, the standard example is rather special, in that both these requirements are fulfilled. However, all of Quinn's obstructions (for the case of trivial fundamental group) are realized in this setting.

Furthermore it is a very good pedagogical example for the general case. In fact the method of ambient plumbing introduced by Kervaire [K] gives an easy proof of the following realization lemma:

<u>Lemma 1:</u> For each element of A(Z) there is a codimension two submanifold, K, of the standard sphere with trivial normal bundle so that Quinn's invariant for $(D^{2n}, S^{2n-1}, K^{2n-2})$ is the chosen element.

<u>Proof:</u> Plumb a manifold in the sphere of the appropriate dimension with intersection pairing given by $B + eB^T$. Then correct the plumbing to the appropriate linking pairing as in Kervaire, [K, Theorem II.3, p. 255].

We now further examine the fibred knots. Because ∂M is a sphere, the sequence for the homology of a book shows that $Id-h_*$ is an isomorphism. Next, by a second application of the sequence, we compute the linking pairing, $L(x,y)$ as the intersection pairing $I(x,(Id-h_*)^{-1}y)$. For the cycle $(Id-h_*)z$ is the difference between "pushing" off in the positive versus the negative direction. The result bounds $z \times I$ in the normal bundle, $P \times I$, to $P \times [1/2]$ and intersects x algebraically in $I(x,z)$ points.

In summary, Quinn's form B is the standard linking form L which satisfies $L - (-1)^{k-1}L^T$ = Intersection Form which is unimodular. A second intersecting corollary of Quinn's method of proof is the following:

<u>Theorem 3:</u> A high dimensional fibred knot is concordant to zero iff it is fibered concordant to zero (that is, the slice disc bounding the knot in D^{2k} can be chosen so that the fibration of the complement of the knot ex-

tends to the complement of the slice disc)

A further result in the case of fibred knots is the simplified relationship between the Seifert linking invariant for a fibred knot and the monodromy of the fibration. Explicitly, denote by $A_e(Z)$ the subgroup of $A(Z)$ with representatives $[N,B]$ whose e-symmetrization, $B + e\,B^T$ is unimodular. Let $C_f^e(Z)$ denote the concordance classes represented by fibred knots and $W_f^e(Z,C_\infty)$ the subgroup of the Witt group of isometries, h, satisfying the condition that (Id -h) be invertible. We then have the following result:

Proposition 3.5: The three groups are isometric: $A_e(Z) \cong C_f^e(Z) \cong W_f^e(Z,C_\infty)$

Proof: By Levine [L2], the linking pairing, L, determines the concordance class of a fibred knot and a fibred knot must satisfy the condition that the linking pairing is a unimodular bilinear pairing (in general only the e-symmetrization need be unimodular.) Furthermore, by the Wang sequence for the fibration of the knot complement, the monodromy must satisfy the above stated condition giving the second isomorphism above.

5) The simple almost Canonical Case

For the general case, we first introduce the setting employed by Frank Quinn. A book decomposition of N is almost canonical if the page P has the homotopy type of an $[n/2]$-complex, where n is the dimension of N. We now return to the setting: M is a 2k-dimensional manifold, ∂M has an almost canonical book decomposition with page P (which then has the homotopy type of a (k-1)-complex) and further we demand that $(M, \partial M)$ is "simple", that is $H_i(M,P\times I)=0$

for i not equal to k. (This terminology was introduced by Levine [L1] in the setting of higher dimensional knot theory.) These restrictions may be satisfied by varying the initial data by a book decomposition bordism [Q, Prop. 11.1, p. 70 and the beginning of the proof of Theorem 6.1]. Note that the book decomposition with empty binding on the mapping torus of a diffeomorphism of a closed manifold is NOT almost canonical! Under the above assumptions, $H_i(M,PxI) = 0$ except for $i=k$, $H_i(PxI) = H_{2k-i-1}(P, \partial P)=0$ for i greater that k-1. Hence the only non-trivial portion of the standard duality diagram is:

$$0 \longrightarrow H_k(M) \longrightarrow H_k(M,PxI) \longrightarrow H_{k-1}(PxI) \longrightarrow H_{k-1}(M) \longrightarrow 0$$

$$0 \longleftarrow H_k(M, \partial M) \longleftarrow H_k(M, \partial M-(PxI)) \longrightarrow H_k(M, \partial M-(PxI)) \longleftarrow H_k(M, \partial M) \longleftarrow 0$$

Let K be the kernel of the inclusion in homology of PxI in M, i.e. those (k-1)-cycles in P which bound in M giving relative cycles in the domain of definition of Quinn's form. Similarly, the dual group $K^*=\mathrm{Hom}_Z(K,Z)$ is an appropriate cokernel. The desired relationship between the two approaches will follow from a detailed analysis of the above duality diagram of extensions.

Now, the intersection pairing on $H_k(M)$ may have a huge radical and, even if we divide out by this abberation, we still may not have a unimodular integral form. This begins the constructions necessary for the proof of Theorem 1. The remainder of the proof will follow the same outline as the identifications made in the standard example.

6) Pasting Lattices and Bilinear Forms

We begin with the following data: J and K are lattices, that is finitely generated free Z-modules equipped with non-degenerate bilinear forms I and L respectively. No symmetry assumption is made on I and L, only that their left (and hence both) adjoint $ad_L I : J \longrightarrow Hom(J,Z)$ is injective. By the standard construction, the quotient $T = Hom(JZ)/ad_L I(J)$ is a finitely generated torsion Z-module which supports a naturally induced Q/Z-valued bilinear linking form, ∂I, induced by extending I to a Q-valued form on the associated vector space. Furthermore, the linking form on the torsion module T is nonsingular (the adjoint ad $\partial I : T \longrightarrow Hom_Z(T,Q/Z)$ is an isomorphism). (See my Memoir [ST, Chapter One and Two] or Alexander, Conner and Hamrick, [ACH] or Walter Neumann's Equivariant Witt Rings [N], for more details.) Again the induced torsion form ∂I need not have any symmetry properties (although it will share any symmetries that the original bilinear form I possesses.) If the two torsion modules are isometric by an isomorphism preserving the forms (an isometry) then, by the standard techniques of localization in the theory of Witt groups of bilinear forms, we may "paste" them together to obtain a unimodular bilinear form over the integers. We will refer to the above procedure as the boundary construction and denote the result $\partial(J,I)$.

Proposition 4: Let (J,I) and (K,L) be two lattices with nondegenerate bilinear forms and an isometry, g, of their associated torsion linking forms. Then the inverse image of the graph of g under $p: J+K \longrightarrow T+T$ supports a unimodular bilinear form induced by I + -L. Furthermore, the inverse image is an extension of the dual of K by J, i.e. $0 \rightarrow J \rightarrow N \rightarrow K^* \rightarrow 0$ is exact.

Proof: Standard, see above references, noting the induced form has the graph of the isometry g as a metabolizer.

We will denote the form induced on N by I + -L by the notation Paste (J,I;K,L). We may generalize this construction slightly by permitting the addition to N of a free module F and its dual F^*, provided with the unimodular e-hyperbolic form $H((x,f),(y,g)) = f(y) + eg(x)$.

7) Computation of Quinn's Invariant

Utilizing the construction of Section 6, we will identify Quinn's form in three steps.

Step 1: First consider the intersection form on M, I restricted to $H_k(M)$. Let F be the radical of I, that is $F = \{x: I(x, H_k(M)) = 0\}$. This is a pure submodule, hence $H_k(M)$ splits into summands F and J. Furthermore the restriction of the intersection form to J is non-degenerate. We note for further reference that F may be identified with the image of $H_k(\partial M)$ in $H_k(M)$ using the homology exact sequence of $(M, \partial M)$ and identifying the adjoint of I as in Hirzebruch's 1962 Berkeley Notes [H]. Utilize the exact sequence a second time to identify T as the submodule of the torsion in $H_{k-1}(\partial M)$ which bound in $H_{k-1}(\partial M)$ provided with the standard Q/Z-valued linking form, lk.

Proposition 5: The boundary construction $\partial(J,I) = (T,-lk)$

Proof: T is also the torsion subgroup of the image from $H_k(M,\partial M)$. The proof is standard (and provided the fundamental geometric interpretation of the boundary in Witt theory.) (e.g. see [AHV, Thm.2.1], or p.70 for a very similar proposition.)

Step 2: Working on the edge.

Let N^* be the kernel of the surjection $j: H_{k-1}(P) \longrightarrow H_{k-1}(M)$ or equivalently the image of the boundary $\partial: H_k(M, P \times I) \longrightarrow H_{k-1}(P)$ in the standard duality diagram. The exact sequence of the book decomposition reduces to:

$$0 \longrightarrow H_{k-1}(P, \partial P)/\mathrm{Ker}(\mathrm{Id} - h_*) \longrightarrow H_{k-1}(P) \longrightarrow H_{k-1}(\partial M) \longrightarrow 0$$

(this uses the fact that P is a (k-1)-complex. Thus, the next term $H_{k-1}(P, \partial P) = H^k(P)$ vanishes.) We now proceed to orthogonally split $H_{k-1}(P)$ into F^*, the inverse image of the free summand of $H_{k-1}(\partial M)$ of cycles of infinite order which bound in M = free part of the kernel of the inclusion $H_{k-1}(\partial M) \longrightarrow H_{k-1}(M)$ and a complementary summand K^*. In the above exact sequence, all that remains is:

$$0 \longrightarrow K = \mathrm{Image}\ (\mathrm{Id} - h_*) \longrightarrow K^* \longrightarrow T \longrightarrow 0$$

Step 3: Identification of the linking form L and its boundary bL.

Viewing K as a submodule of $H_{k-1}(P)$ and observing, as above, that K = Image $(\mathrm{Id}-h^*)$, we may define a bilinear form L on K by $L(x,y) = \mathrm{Int}_p(x, (\mathrm{Id}-h^*)^{-1}y)$. This is well-defined since two liftings differ by an element in the kernel of $\mathrm{Id}-h^*$ which is orthogonal to K. (The kernel is non-singularly paired with F^* under the intersection pairing between the free parts of $H_k(\partial M)$ and $H_{k-1}(\partial M)$. As in the case of fibred knots, the form L agrees with that induced by Quinn's form B. It remains only to check that the linking form induced on T by the boundary construction for L is the standard linking form of $H_{k-1}(\partial M)$ (and the negative of that induced by the intersection on J.)

Proposition 6: $\partial(K,L) = (T,lk)$

Proof: The sequence above $0 \longrightarrow K \longrightarrow K^* \longrightarrow T \longrightarrow 0$ is the one used in the construction of $\partial(K,L)$. If x and y are in K^* representing classes in T and ny is in the image of K for some integer n, then L(x,ny) can be viewed, as in the section on fibred knots, as the intersection of x and a cycle in ∂M bounding ny. But dividing this intersection number by n and viewing the result in Q/Z is precisely the torsion linking, lk(x,y).

Finally, one must check that the pasting construction gives Quinn's form B but all the necessary details to obtain this result from the standard duality diagram have already been revealed in the above exposition.

Theorem 3: The invariant of Quinn $(H_k(M,P{\times}I),B)$ is equal to Paste $(J+F,I;K+F^*,L)$ which is Witt-equivalent (even in the hyperbolic sense) to Paste (J,I;K,L).

8) Verification of the Algebraic Relationship

We now construct the algebraic counterpart of the geometric doubling $D(P) = [P \cup P, h|_P \cup Id]$

The algebraic analog of the doubling construction involves a map which we will call bilinearization and denote bil: $A(R) \longrightarrow W^e(R,C_\infty)$. The prototype for this map is the well known decomposition for any matrix B (over any ring in which 2 is a unit) as a sum of a symmetric and a skew symmetric matrix, $B = (B+B^T)/2 + (B-B^T)/2$. Less well known is the fact (used by Levine [L2] in his computation of knot concordance, that $S = (-e)B^{-1}B^T$ is an isometry of the above forms, $Q = B + eB^T$ defined by the splitting of B:

$$S^TQS = (-e)BB^{-T}(B+eB^T)(-e)B^{-1}B^T = BB^{-T}B^T + eBB^{-1}B^T = B + eB^T$$

<u>Remark:</u> If $B+eB^T$ is unimodular, then $[Q,S]$ determines a well-defined element of $W^e(Z,C_\infty)$. Conversely, as utilized in Prop. 3.5, if the isometry satisfies the condition that (Id - S) is invertible, the triple $[M,Q,S]$ determines the element $[M,B = Q(Id-S)^{-1}]$.

<u>Proposition 7:</u> There is a homomorphism bil: $A(R) \longrightarrow W(R,C_\infty)$ extending the e-bilinearization map from the subgroup $A_e(R) = \{B: B+eB^T \text{ is unimodular}\}$. The map is explicitly given by:

$$\text{bil}(N,B) = \text{Paste}(K=N/\text{Radical}(B+eB^T),\ B+eB^T,\ -eB^{-1}B^T;K,\ -(B+eB^T),\ \text{Id}).$$

<u>Proof:</u> As B is a unimodular bilinear pairing, we may define an a-symmetry operator, denoted h, which is an isomorphism from N to N^* relating the left and right adjoints of B by the formula $B(x,-eh(y)) = B(y,x)$. (The e is introduced to relate properly with the monodromy in the case of a fibred knot and for sign conventions to be satisfied later in the proof.)

Next we consider the e-symmetrization of B, $S(x,y) = B(x,y) + eB(y,x) = B(x,(I-h)y)$. The radical of S, $R = \{x:S(N,y) = 0\}$ = Kernel (I-h), is a pure submodule of N on which S is identically zero (and on which B is (-e)-symmetric.) S induces a non-degenerate (but not necessarily unimodular) e-symmetric bilinear form on K = N/R. Furthermore h is an isometry of S for:

$$S(hx,hy) = B(hx,hy)+eB(hy,hx) = B(y,eh(x))+eB(x,eh(y)) = B(x,y)+eB(y,x)$$

To compute the boundary of (K,S,h) we recall $S(x,y) = B(x,(I-h)y)$ which implies that the image of the adjoint of S in the dual of K,$K^{\#}$, is the image of (I-h): $K^{\#} \longrightarrow K^{\#}$. Therefore we have isomorphisms, $K^{\#}/\text{ad } S(N) = K^{\#}/(Id-h)(K^{\#})$ and it follows that h induces the identity on the torsion module T (the above quotient) which occurs in the boundary construction on (K,S,h). Hence (K,S,id) has the same boundary and the Paste construction Paste(K,S,h;K,-S,id) gives a well-defined element of $W^e(R,C_\infty)$.

This completes the series of definitions necessary in algebraically tracking down the relation between Quinn's obstruction and the isometric structure.

Remark: Neither the unimodularity nor even the non-degeneracy of the bilinear form B was utilized in the definition of bil $[N,B]$ via the Paste construction.

Example 1: We now give an explicit example, in terms of integral matrices, of the bilinearization homomorphism for the case e = +1. Suppose B is the form with the matrix below on a rank four free Z-module.

$$B = \begin{bmatrix} 1 & 1 & 0 & 0 \\ 0 & 1 & 0 & 0 \\ 0 & 0 & 0 & 1 \\ 0 & 0 & -1 & 0 \end{bmatrix}$$

Then the radical of the +1-symmetrization of B, $B+B^T$ has rank two and can be viewed as the Z-module spanned by the last two elements in the chosen ordered basis. Therefore $N/R = K$ is the rank two free Z-module provided with the form $\begin{bmatrix} 2 & 1 \\ 1 & 2 \end{bmatrix}$

The boundary of this is the torsion Z-module, Z/(3). Pulling back the inverse image of the diagonal, as outlined above, we obtain a rank four representative of the image of bil $[N,B]$ with matrix representation:

$$S = \begin{bmatrix} 0 & 0 & 1 & 0 \\ 0 & 0 & 0 & -1 \\ 1 & 0 & 2 & 0 \\ 0 & -1 & 0 & -2 \end{bmatrix} \qquad h = \begin{bmatrix} 1 & -1 & -1 & 0 \\ 0 & -1 & -2 & 0 \\ -1 & 1 & 0 & 0 \\ 0 & 1 & 1 & 0 \end{bmatrix}$$

We now prove theorem 1:

<u>Proof (Theorem 1):</u> The commutativity of the part of the diagram involving the invariants on closed manifolds is easily verified. The only non-trivial verification involves the commutativity of the diagram:

$$\begin{array}{ccc} BB^{SO}_{2n}(X;Z,1) & \longrightarrow & \Delta_{2n} \\ \downarrow & & \downarrow \\ A(R=Z[\pi_1(X)]) & \longrightarrow & W^e(R,C\ \ ) \end{array}$$

First we observe that the geometric homomorphism on the top line is given by the doubling construction on the monodromy of the page:

$$D(P,h) = (P \cup P; h \cup Id_{P})$$

It will suffice to compute the isometric structure of this object and compare it with Quinn's invariant via the bilinearization map. From the Mayer-Vietoris exact sequence for the double it follows that I(D(P,h)) is given by: Paste ($G = H_n(P)$/Radical (Int_P), Int_P, h_*;L,-Int_P,Id) where Int_P is the intersection pairing on $H_n(P)$. From Section Seven we have that Quinn's invariant is given by $[N,B]$ = Paste($J=H_k(M)$/Radical(Int_M), Int_M; $K = Image(Id-h_*):H_{k-1}(P) \longrightarrow H_{k-1}(P)$, $L(x,y) = Int_P(x,(Id-h_*)^{-1}(y))$. Now the intersection pairing on M is $(-1)^k = (-e)$-symmetric hence J Radical of B! Therefore bil $[N,B]$ = bil $[K,L]$ (in an extended sense of the remark, for L is not unimodular.) The last step in our verification is to check that bil $[K,L]$ = Paste (G,Int_P,h_*;G,-Int_P,Id). In fact as the definition of bil is the paste construction on (K/Radical ($L+eL^T$), $S=L+eL^T$, $-eL^{-1}L^T$; K/Radical, -S,Id) it suffices to show that J is isomorphic with K/Radical $L+eL^T$ so that Int_P is identified appropriately with the bilinear form $L = Int_P(x, (Id-h_*)^{-1}y)$. But in Prop. 7 we demonstrated that K = Image (Id -h) = $H_{k-1}(P)/Ker(Id-h_*)$ = K/Radical(Int_P). Finally we verify that the forms are identified appropriately. But we have the following relationships:

$I(y,(Id-h)^{-1}x) = eI((Id-h)^{-1}x,y) = eI(x,(Id-h^{-1})^{-1}y)$ (since h is an isometry of the intersection pairing on P) $= I(x,-eh(Id-h)^{-1}y)$ and -eh is the asymmetry operator for the bilinear form L above. This gives the desired relationship, $(L=eL^T)(x,y) = I(x,(Id-h)^{-1}y) + eI(x,-eh(Id-h)^{-1}y) = I(x,y)$, over the rational field, hence integrally as well, yielding the desired conclusion.

Remark: The torsion linking form on the boundary of $M, T =$ Torsion $(H_{k-1}(t(P,h), lk)$ has been resolved in two distinct ways: First by $H_k(M)$ and the (-e)-symmetric intersection form on M which combines to give the invariant in Quinn's group. In the second resolution the middle dimensional homology of M is replaced by that of P giving the relationship determined by the composition of the doubling operation and the bilinearization map.

9) The de Rham Invariant and further Remarks

In this section we will complete the verification of the exactness of the algebraic exact sequence (for the case R=Z). First we recall the localization sequence:

$$0 \longrightarrow W^e(Z) \longrightarrow W^e(Q) \xrightarrow{\partial_e} W^e(Q/Z) \longrightarrow \text{Coker } \partial_e \longrightarrow 0$$

Example 2: The cokernel is trivial for e = +1 by a theorem of Milnor-Tate (with roots in work of Gauss). However in the case e = -1 the cokernel is Z/2 with the non-trivial element given by the class of the skew-symmetric form (in characteristic two) $(Z/2, \langle 1 \rangle)$, and is detected by the de Rham invariant.

The definition of the map $W^e(R, C_\infty) \longrightarrow \text{Coker } \partial_{-e}$ is given as follows: Let $[M,S,h]$ be a representative of the given class and denote by L the lattice, Image (Id-h). Define the bilinear form, B(x,y) on L by $B(x,y) = S(x,(Id-h)^{-1}y))$. Let the boundary of (L,B) be denoted (T,b).

Claim: The torsion bilinear form b is (-e)-symmetric.

This follows from computing the symmetry operator of B, $B(y,x)=B(x,-eh^{-1}y)$ and the previous note that the torsion module in the boundary construction is isomorphic to the torsion submodule of $M/(Id-h)(M)$. Therefore h induces the identity on T. Combining these facts verifies the claim.
The second component of the final map in the algebraic exact sequence maps $[M,S,h]$ to the class of (T,b) in the cokernel of the boundary in the localization sequence above. We now give the proof of Theorem 2:

Proof (Theorem 2): The injectivity of the first map is trivial by the transistivity of the metabolic equivalence relation and the definition of the map as a forgetful map. Next we investigate the exactness at A(Z).

First we note the underlying secret: The objects $[N,B]$ in A(Z) and bil $[N,B]$ in $W(Z,C_\infty)$ differ only in the eigenspace of eigenvalue one, for we past N modulo the radical (that is the kernel of Id - h) together with the identity as an isometry. If bil $[N,B]$ is metabolic then the eigenvalue one subspace is metabolic and hence N/Radical S is metabolic. But this means that N is Witt equivalence in A(Z) to a finite extension of the Radical of $(B+eB^T)$. But for any such space the restriction of the bilinear form B is (-e)-symmetric and hence $[N,B]$ is in the image from the (-e)-symmetric Witt group.

It remains to show surjectivity of the final map. It is trivial to find a pre-image for the first component, for example, $[L,S,Id]$ maps to $[L,S]$ under the forgetful map. For the second component (in the sigle non-trivial case) we observe that the class of $[Z, \langle 1\rangle, -Id] + [Z, \langle -1\rangle, Id]$ is a form with signature zero (and hence trivial in the symmetric Witt group of Z) and de Rham invariant under the construction given above. Therefore the two invariants can be realized independently and the final map is surjective

The de Rham Invariant

We wish to demonstrate the following proposition which is necessary for the completion of the geometric study of the bordism of diffeomorphisms.

Proposition 8: The de Rham invariant of the mapping torus of a diffeomorphism with isometric structure I(P,h) is determined by the map $W^1(Z,C_\infty) \longrightarrow Z/2 = \text{Coker}\,\partial$ described above. Explicitly, we have the following relationship. If the isometric structure = $[M,S,h]$, then the de Rham invariant of the mapping torus = $\text{Rank}_Z(\text{Id-h})-\text{Rank}_{F_2}(\text{Id-h/2h})$ modulo two.

Here, h/2h denotes the map induced by h on the mod two reduction of M, M/2M.

Proof: To identify the de Rham invariant of the mapping torus, we proceed as follows: Consider the Wang exact sequence for M_f

$$H_k(M) \xrightarrow{\text{Id - } h_*} H_k(M) \longrightarrow H_k(M_h) \longrightarrow H_{k-1}(M) \longrightarrow$$

Now the element determined by the algebraic Z/2 invariant on the Witt group of isometries corresponds to the cokernel of $(\text{Id-}h_*)$ restricted to the torsion free subgroup of $H_k(M)$. The de Rham invariant is the mod two rank of the torsion subgroup of $H_k(M_h)$ tensored with Z/2 and we have identified the contribution arising from the cokernel of (Id-h) on the free part. The remainder of the torsion in the (k-1)-homology group of the mapping torus arises from the cokernel of (Id-h) restricted to the torsion subgroup of

$H_k(M_h)$ and the kernel of (Id-h) in dimension (k-1). But, Poincare duality gives a non-singular pairing between these groups into Q/Z:

Tor $H_k(M)$ x Tor $H_{k-1}(M) \longrightarrow Q/Z$. But the cokernel of (Id-h) restricted to Tor $H_k(M)$ is a self-annihilating subspace under the pairing into Q/Z, hence $[\text{Tor } H_k(M_h), 1k]$ is Witt equivalent to T = Coker (Id-h) restricted to the free part of $H_k(M)$.

Even Forms

As a corollary of our work on the algebraic exact sequence of Theorem 2, we reprove the following result on G. Hamrick on the subgroup of $W(Z,C_\infty)$ admitting even representatives. This is of geometrical interest in the study of diffeomorphisms preserving further geometric structures (in particular Spin structures.)

Corollary 2: There is an exact sequence:

$$0 \longrightarrow W_{ev}(Z,C_\infty) \longrightarrow W(Z,C_\infty) \longrightarrow Z/8 + Z/2 \longrightarrow 0.$$

The map to Z/8 is given by the signature of the underlying symmetric form taken modulo eight and the second component is the de Rham invariant.

Proof: Apply the algebraic exact sequence of Theorem two noting that the image of the bilinearization map supports even forms. If the signature is zero modulo eight but non-zero add an appropriate number of copies of the even eight dimensional form of index eight (or its negative) provided with the identity isometry. This is then in the image of the bilinearization map. Finally we recover the original form, up to Witt equivalence, by subtracting what we had previously added.

The relation between the Witt group of isometries and bilinear forms was first observed by Levine (with the aide of Milnor) in his study of the knot concondance group, but he did this only over the rational field. The refinements above necessary to handle the integral questions are very similar to the techniques of my Memoir. A student of Pierre Conner, Max Warshauer, has made a study of Quinn's group, denoting it A(Z) (asymmetric), and he has given a complete computation following the methods of my Memoir [St].

The generalization of the above construction to the case when R is the group ring of a non-trivial fundamental group can be made in many situations where there is a localization sequence for that ring (see e.g. [P]). For simplicity of exposition, we have constrained our proof to the case R = Z.

Relevant Bibliography

[AHV] Alexander, J. Hamrick, G. & Vick, J. Linking Forms and Maps of odd prime order, Trans. AMS, 221,(1976), p.169-185.

[ACH] Alexander, J. Conner, P.E.& Hamrick, G. Odd Order Group Actions and Classification of Inner Products, Lecture Notes in Mathematics #625, Springer, 1977.

[CF] Conner, P.E. Floyd, E.E. Differentiable Periodic Maps, Springer, 1964. (Second edition, LNIM # 738, 1979)

[H] Hirzebruch, F. Differentiable Manifolds and Quadratic Forms, (Notes by S.S. Koh, edited by D.W. Neumann, with an Appendix by W. Scharlau), M.Dekker,1971.

[K] Kervaire, M. Les noeuds de dimensions superieures, Bull. Soc. Math. de France 93 (1965) p. 225-271.

[Kr] Kreck, M, Bordism of Diffeomorphisms, Bull AMS 82 (1976), 779 - 61.

[L1] Levine, J. Unknotting spheres in codimension two, Topology 4 (1965), p. 9-16.

[L2] Levine, J. Knot cobordism in codimension two, Comm.Math. Helv. 44 (1968), p. 229-244.

[LdM] Lopez de Medrano, S. Cobordism of diffeomorphisms of (k-1)-connected 2k-manifolds, Second Conference on Compact Transformation Groups, pp. 217-227, LNIM # 298, Springer, 1972.

[MH] Milnor, J. & Husemoller, D. Symmetric Bilinear Forms, Springer (1972).

[M2] Milnor, J. Infinite Cyclic Coverings, Conference on the Topology of Manifolds, ed. John G. Hocking. Prindle, Weber and Schmidt (1968), p. 115-34.

[N] Neumann, D.W. Equivariant Witt Rings, Bonner Mathematische Schriften Nr. 100 (Bonn, 1977/78).

[P] Pardon, W. The exact sequence of a localization for Witt groups, Lecture Notes in Mathematics # 551 (1976) p. 336 - 79.

[Q] Quinn, F. Open book decompositions and the bordism of automorphisms, Topology 18 (1979), p. 55-73.

[R] de Rham, G. Sur L'Analysis Situs, J. Math. Pures Appl., 10 (1931).

[St] Stoltzfus, N.W. Unraveling the integral knot concordance group. Mem. AMS 192 (1976).

[W] Winkelnkemper,E. On equators of manifolds. (Thesis Princeton Univ., 1969).

[War] Warshauer, M. The Witt group of degree k maps and asymmetric inner product spaces, LNIM # 914, Springer , 1982.

SUBJECT INDEX

ol. 926: Geometric Techniques in Gauge Theories. Proceedings, 981. Edited by R. Martini and E.M.de Jager. IX, 219 pages. 1982.

ol. 927: Y. Z. Flicker, The Trace Formula and Base Change for L (3). XII, 204 pages. 1982.

ol. 928: Probability Measures on Groups. Proceedings 1981. Edited y H. Heyer. X, 477 pages. 1982.

ol. 929: Ecole d'Eté de Probabilités de Saint-Flour X – 1980. roceedings, 1980. Edited by P.L. Hennequin. X, 313 pages. 1982.

ol. 930: P. Berthelot, L. Breen, et W. Messing, Théorie de Dieudonné ristalline II. XI, 261 pages. 1982.

ol. 931: D.M. Arnold, Finite Rank Torsion Free Abelian Groups nd Rings. VII, 191 pages. 1982.

ol. 932: Analytic Theory of Continued Fractions. Proceedings, 1981. dited by W.B. Jones, W.J. Thron, and H. Waadeland. VI, 240 pages. 982.

ol. 933: Lie Algebras and Related Topics. Proceedings, 1981. dited by D. Winter. VI, 236 pages. 1982.

ol. 934: M. Sakai, Quadrature Domains. IV, 133 pages. 1982.

ol. 935: R. Sot, Simple Morphisms in Algebraic Geometry. IV, 6 pages. 1982.

ol. 936: S.M. Khaleelulla, Counterexamples in Topological Vector paces. XXI, 179 pages. 1982.

ol. 937: E. Combet, Intégrales Exponentielles. VIII, 114 pages. 982.

ol. 938: Number Theory. Proceedings, 1981. Edited by K. Alladi. , 177 pages. 1982.

ol. 939: Martingale Theory in Harmonic Analysis and Banach paces. Proceedings, 1981. Edited by J.-A. Chao and W.A. Woyzyński. VIII, 225 pages. 1982.

ol. 940: S. Shelah, Proper Forcing. XXIX, 496 pages. 1982.

ol. 941: A. Legrand, Homotopie des Espaces de Sections. VII, 2 pages. 1982.

ol. 942: Theory and Applications of Singular Perturbations. Proeedings, 1981. Edited by W. Eckhaus and E.M. de Jager. V, 363 ages. 1982.

ol. 943: V. Ancona, G. Tomassini, Modifications Analytiques. IV, 0 pages. 1982.

ol. 944: Representations of Algebras. Workshop Proceedings, 980. Edited by M. Auslander and E. Lluis. V, 258 pages. 1982.

l. 945: Measure Theory. Oberwolfach 1981, Proceedings. Edited D. Kölzow and D. Maharam-Stone. XV, 431 pages. 1982.

l. 946: N. Spaltenstein, Classes Unipotentes et Sous-groupes de rel. IX, 259 pages. 1982.

l. 947: Algebraic Threefolds. Proceedings, 1981. Edited by Conte. VII, 315 pages. 1982.

l. 948: Functional Analysis. Proceedings, 1981. Edited by D. Butvić, H. Kraljević, and S. Kurepa. X, 239 pages. 1982.

l. 949: Harmonic Maps. Proceedings, 1980. Edited by R.J. Knill, Kalka and H.C.J. Sealey. V, 158 pages. 1982.

l. 950: Complex Analysis. Proceedings, 1980. Edited by J. Eells. 428 pages. 1982.

l. 951: Advances in Non-Commutative Ring Theory. Proceedings, 81. Edited by P.J. Fleury. V, 142 pages. 1982.

l. 952: Combinatorial Mathematics IX. Proceedings, 1981. Edited E. Billington, S. Oates-Williams, and A.P. Street. XI, 443 pages. 82.

Vol. 953: Iterative Solution of Nonlinear Systems of Equations. Proceedings, 1982. Edited by R. Ansorge, Th. Meis, and W. Törnig. VII, 202 pages. 1982.

Vol. 954: S.G. Pandit, S.G. Deo, Differential Systems Involving Impulses. VII, 102 pages. 1982.

Vol. 955: G. Gierz, Bundles of Topological Vector Spaces and Their Duality. IV, 296 pages. 1982.

Vol. 956: Group Actions and Vector Fields. Proceedings, 1981. Edited by J.B. Carrell. V, 144 pages. 1982.

Vol. 957: Differential Equations. Proceedings, 1981. Edited by D.G. de Figueiredo. VIII, 301 pages. 1982.

Vol. 958: F.R. Beyl, J. Tappe, Group Extensions, Representations, and the Schur Multiplicator. IV, 278 pages. 1982.

Vol. 959: Géométrie Algébrique Réelle et Formes Quadratiques, Proceedings, 1981. Edité par J.-L. Colliot-Thélène, M. Coste, L. Mahé, et M.-F. Roy. X, 458 pages. 1982.

Vol. 960: Multigrid Methods. Proceedings, 1981. Edited by W. Hackbusch and U. Trottenberg. VII, 652 pages. 1982.

Vol. 961: Algebraic Geometry. Proceedings, 1981. Edited by J.M. Aroca, R. Buchweitz, M. Giusti, and M. Merle. X, 500 pages. 1982.

Vol. 962: Category Theory. Proceedings, 1981. Edited by K.H. Kamps, D. Pumplün, and W. Tholen, XV, 322 pages. 1982.

Vol. 963: R. Nottrot, Optimal Processes on Manifolds. VI, 124 pages. 1982.

Vol. 964: Ordinary and Partial Differential Equations. Proceedings, 1982. Edited by W.N. Everitt and B.D. Sleeman. XVIII, 726 pages. 1982.

Vol. 965: Topics in Numerical Analysis. Proceedings, 1981. Edited by P.R. Turner. IX, 202 pages. 1982.

Vol. 966: Algebraic K-Theory. Proceedings, 1980, Part I. Edited by R.K. Dennis. VIII, 407 pages. 1982.

Vol. 967: Algebraic K-Theory. Proceedings, 1980. Part II. VIII, 409 pages. 1982.

Vol. 968: Numerical Integration of Differential Equations and Large Linear Systems. Proceedings, 1980. Edited by J. Hinze. VI, 412 pages. 1982.

Vol. 969: Combinatorial Theory. Proceedings, 1982. Edited by D. Jungnickel and K. Vedder. V, 326 pages. 1982.

Vol. 970: Twistor Geometry and Non-Linear Systems. Proceedings, 1980. Edited by H.-D. Doebner and T.D. Palev. V, 216 pages. 1982.

Vol. 971: Kleinian Groups and Related Topics. Proceedings, 1981. Edited by D.M. Gallo and R.M. Porter. V, 117 pages. 1983.

Vol. 972: Nonlinear Filtering and Stochastic Control. Proceedings, 1981. Edited by S.K. Mitter and A. Moro. VIII, 297 pages. 1983.

Vol. 973: Matrix Pencils. Proceedings, 1982. Edited by B. Kågström and A. Ruhe. XI, 293 pages. 1983.

Vol. 974: A. Draux, Polynômes Orthogonaux Formels – Applications. VI, 625 pages. 1983.

Vol. 975: Radical Banach Algebras and Automatic Continuity. Proceedings, 1981. Edited by J.M. Bachar, W.G. Bade, P.C. Curtis Jr., H.G. Dales and M.P. Thomas. VIII, 470 pages. 1983.

Vol. 976: X. Fernique, P.W. Millar, D.W. Stroock, M. Weber, Ecole d'Été de Probabilités de Saint-Flour XI – 1981. Edited by P.L. Hennequin. XI, 465 pages. 1983.

Vol. 977: T. Parthasarathy, On Global Univalence Theorems. VIII, 106 pages. 1983.

Vol. 978: J. Ławrynowicz, J. Krzyż, Quasiconformal Mappings in the Plane. VI, 177 pages. 1983.

Vol. 979: Mathematical Theories of Optimization. Proceedings, 1981. Edited by J.P. Cecconi and T. Zolezzi. V, 268 pages. 1983.

Vol. 980: L. Breen. Fonctions thêta et théorème du cube. XIII, 115 pages. 1983.

Vol. 981: Value Distribution Theory. Proceedings, 1981. Edited by I. Laine and S. Rickman. VIII, 245 pages. 1983.

Vol. 982: Stability Problems for Stochastic Models. Proceedings, 1982. Edited by V. V. Kalashnikov and V. M. Zolotarev. XVII, 295 pages. 1983.

Vol. 983: Nonstandard Analysis-Recent Developments. Edited by A. E. Hurd. V, 213 pages. 1983.

Vol. 984: A. Bove, J. E. Lewis, C. Parenti, Propagation of Singularities for Fuchsian Operators. IV, 161 pages. 1983.

Vol. 985: Asymptotic Analysis II. Edited by F. Verhulst. VI, 497 pages. 1983.

Vol. 986: Séminaire de Probabilités XVII 1981/82. Proceedings. Edited by J. Azéma and M. Yor. V, 512 pages. 1983.

Vol. 987: C. J. Bushnell, A. Fröhlich, Gauss Sums and p-adic Division Algebras. XI, 187 pages. 1983.

Vol. 988: J. Schwermer, Kohomologie arithmetisch definierter Gruppen und Eisensteinreihen. III, 170 pages. 1983.

Vol. 989: A. B. Mingarelli, Volterra-Stieltjes Integral Equations and Generalized Ordinary Differential Expressions. XIV, 318 pages. 1983.

Vol. 990: Probability in Banach Spaces IV. Proceedings, 1982. Edited by A. Beck and K. Jacobs. V, 234 pages. 1983.

Vol. 991: Banach Space Theory and its Applications. Proceedings, 1981. Edited by A. Pietsch, N. Popa and I. Singer. X, 302 pages. 1983.

Vol. 992: Harmonic Analysis, Proceedings, 1982. Edited by G. Mauceri, F. Ricci and G. Weiss. X, 449 pages. 1983.

Vol. 993: R. D. Bourgin, Geometric Aspects of Convex Sets with the Radon-Nikodým Property. XII, 474 pages. 1983.

Vol. 994: J.-L. Journé, Calderón-Zygmund Operators, Pseudo-Differential Operators and the Cauchy Integral of Calderón. VI, 129 pages. 1983.

Vol. 995: Banach Spaces, Harmonic Analysis, and Probability Theory. Proceedings, 1980–1981. Edited by R.C. Blei and S.J. Sidney. V, 173 pages. 1983.

Vol. 996: Invariant Theory. Proceedings, 1982. Edited by F. Gherardelli. V, 159 pages. 1983.

Vol. 997: Algebraic Geometry – Open Problems. Edited by C. Ciliberto, F. Ghione and F. Orecchia. VIII, 411 pages. 1983.

Vol. 998: Recent Developments in the Algebraic, Analytical, and Topological Theory of Semigroups. Proceedings, 1981. Edited by K. H. Hofmann, H. Jürgensen and H. J. Weinert. VI, 486 pages. 1983.

Vol. 999: C. Preston, Iterates of Maps on an Interval. VII, 205 pages. 1983.

Vol. 1000: H. Hopf, Differential Geometry in the Large, VII, 184 pages. 1983.

Vol. 1001: D.A. Hejhal, The Selberg Trace Formula for PSL(2, ℝ). Volume 2. VIII, 806 pages. 1983.

Vol. 1002: A. Edrei, E.B. Saff, R.S. Varga, Zeros of Sections of Power Series. VIII, 115 pages. 1983.

Vol. 1003: J. Schmets, Spaces of Vector-Valued Continuous Functions. VI, 117 pages. 1983.

Vol. 1004: Universal Algebra and Lattice Theory. Proceedings, 1982. Edited by R. S. Freese and O. C. Garcia. VI, 308 pages. 1983.

Vol. 1005: Numerical Methods. Proceedings, 1982. Edited by V. Pereyra and A. Reinoza. V, 296 pages. 1983.

Vol. 1006: Abelian Group Theory. Proceedings, 1982/83. Edited by R. Göbel, L. Lady and A. Mader. XVI, 771 pages. 1983.

Vol. 1007: Geometric Dynamics. Proceedings, 1981. Edited by J. Palis Jr. IX, 827 pages. 1983.

Vol. 1008: Algebraic Geometry. Proceedings, 1981. Edited by J. Dolgachev. V, 138 pages. 1983.

Vol. 1009: T. A. Chapman, Controlled Simple Homotopy Theory an Applications. III, 94 pages. 1983.

Vol. 1010: J.-E. Dies, Chaînes de Markov sur les permutations. I 226 pages. 1983.

Vol. 1011: J.M. Sigal. Scattering Theory for Many-Body Quantur Mechanical Systems. IV, 132 pages. 1983.

Vol. 1012: S. Kantorovitz, Spectral Theory of Banach Space Ope ators. V, 179 pages. 1983.

Vol. 1013: Complex Analysis – Fifth Romanian-Finnish Semina Part 1. Proceedings, 1981. Edited by C. Andreian Cazacu, N. Boboc M. Jurchescu and I. Suciu. XX, 393 pages. 1983.

Vol. 1014: Complex Analysis – Fifth Romanian-Finnish Semina Part 2. Proceedings, 1981. Edited by C. Andreian Cazacu, N. Boboc M. Jurchescu and I. Suciu. XX, 334 pages. 1983.

Vol. 1015: Equations différentielles et systèmes de Pfaff dans l champ complexe – II. Seminar. Edited by R. Gérard et J. P. Rami V, 411 pages. 1983.

Vol. 1016: Algebraic Geometry. Proceedings, 1982. Edited by M Raynaud and T. Shioda. VIII, 528 pages. 1983.

Vol. 1017: Equadiff 82. Proceedings, 1982. Edited by H. W. Knobloc and K. Schmitt. XXIII, 666 pages. 1983.

Vol. 1018: Graph Theory, Łagów 1981. Proceedings, 1981. Edited b M. Borowiecki, J. W. Kennedy and M. M. Sysło. X, 289 pages. 198

Vol. 1019: Cabal Seminar 79–81. Proceedings, 1979–81. Edited b A. S. Kechris, D. A. Martin and Y. N. Moschovakis. V, 284 pages. 198

Vol. 1020: Non Commutative Harmonic Analysis and Lie Group Proceedings, 1982. Edited by J. Carmona and M. Vergne. V, 18 pages. 1983.

Vol. 1021: Probability Theory and Mathematical Statistics. Pr ceedings, 1982. Edited by K. Itô and J.V. Prokhorov. VIII, 747 page 1983.

Vol. 1022: G. Gentili, S. Salamon and J.-P. Vigué. Geometry Semina "Luigi Bianchi", 1982. Edited by E. Vesentini. VI, 177 pages. 1983.

Vol. 1023: S. McAdam, Asymptotic Prime Divisors. IX, 118 page 1983.

Vol. 1024: Lie Group Representations I. Proceedings, 1982–198 Edited by R. Herb, R. Lipsman and J. Rosenberg. IX, 369 pages. 198

Vol. 1025: D. Tanré, Homotopie Rationnelle: Modèles de Che Quillen, Sullivan. X, 211 pages. 1983.

Vol. 1026: W. Plesken, Group Rings of Finite Groups Over p-ad Integers. V, 151 pages. 1983.

Vol. 1027: M. Hasumi, Hardy Classes on Infinitely Connected Ri mann Surfaces. XII, 280 pages. 1983.

Vol. 1028: Séminaire d'Analyse P. Lelong – P. Dolbeault – H. Skod Années 1981/1983. Edité par P. Lelong, P. Dolbeault et H. Skod VIII, 328 pages. 1983.

Vol. 1029: Séminaire d'Algèbre Paul Dubreil et Marie-Paule Malliavi Proceedings, 1982. Edité par M.-P. Malliavin. V, 339 pages. 1983.

Vol. 1030: U. Christian, Selberg's Zeta-, L-, and Eisensteinserie XII, 196 pages. 1983.

Vol. 1031: Dynamics and Processes. Proceedings, 1981. Edited b Ph. Blanchard and L. Streit. IX, 213 pages. 1983.

Vol. 1032: Ordinary Differential Equations and Operators. Pr ceedings, 1982. Edited by W. N. Everitt and R. T. Lewis. XV, 521 page 1983.

Vol. 1033: Measure Theory and its Applications. Proceedings, 198 Edited by J. M. Belley, J. Dubois and P. Morales. XV, 317 pages. 198

Vol. 1034: J. Musielak, Orlicz Spaces and Modular Spaces. 222 pages. 1983.